Plants for people

Man picking medicinal plants in Ecuadorian herb garden

Anna Lewington

Plants for people

Foreword by Anita Roddick

NATURAL HISTORY MUSEUM PUBLICATIONS

Royal Botanic Gardens, Kew
The World Wide Fund For Nature (UK) Education Department

To Eddie for his invaluable advice and practical support
and my nephew Aidan,
who patiently shared his Auntie with this book

Designed by Gillian Greenwood

© Anna Lewington 1990
Published by The Natural History Museum
Cromwell Road, London SW7 5BD

British Library Cataloguing in Publication Data
Lewington, Anna
  Plants for people
  1. Plants. Use by man
  I. Title
  630

ISBN 0-565-01094-8

Typeset in Palatino by Precise Printing & Communications Ltd
Bartholomew Press, Leatherhead, Surrey
Printed in Singapore by Times Offset Private Ltd.

Contents

Foreword vi

Introduction vii

Starting the day 1

Keeping us covered 35

From first foods to fast foods 67

House and home — plants that protect us 105

Your very good health — plants that cure us 135

Getting around — plants that transport us 169

Recreation — plants that entertain us 195

Index 224

Author's acknowledgements 230

Picture credits 231

Foreword

The relationship between plants and people is central and fundamental to how we live. I'm delighted with this book — it explores that relationship and draws attention to it. We need plants: for basic human activities like washing and eating, for medicine, and for less obvious reasons like shelter, amusement and transport — plants are an integral part of our lives. I hope that when people read this book they will gain a renewed appreciation of what plants do for us — and will then wonder what *we* can do for the natural world. I hope too that they will realize that we need to care for and nurture plants and trees, and treat them with respect, for it's quite clear we are failing to do so: the human race is destroying the natural world, abusing it, rather than treating it carefully and responsibly. Plants are under threat the world over — from the hedgerows of Britain to the tropical rainforests of the Amazon.

Precious plants (some of them still unknown and unresearched) are being destroyed in the name of the false gods of progress and development — or in the cause of straightforward greed. As natural habitats are destroyed, so are other kinds of life: animals and people. Ancient and still thriving cultures who depend on the forests, who care for the natural world and who know how to live in harmony with it, are under threat. We can learn so much from other peoples such as the Kayapo of Brazil and the Penan of Malaysia, who possess unique expertise and know how to use plants with respect and how to cultivate the land and forests without harming them. Their relationship with the world around them is a subtle one that has existed for centuries, and their wisdom is unique and invaluable.

The North American Indians, in common with many other indigenous peoples, revere the earth as a mother: the sacred provider of life and all it contains. As Bedagi (Big Thunder) of the Wabanakis Nation said in 1900:

'. . . the earth is our mother. She nourishes us; that which we put into the ground she returns to us, and healing plants she gives us likewise . . .'

Let's learn from those beliefs. Yes, plants are for people, but let's also think what people can do for plants and all the natural world while we still have the chance, and before it's too late!

Anita Roddick
MANAGING DIRECTOR, THE BODY SHOP INTERNATIONAL PLC

Introduction

Early man evolved with plants and soon began to use them for food, shelter, clothing and fire. Ever since he has been exploring new ways to use the 250 000 species of flowering plants that surround him. Although we now live in a highly industrialized society, we have not lost this dependence on plants in the least. We may use them in more sophisticated ways to lubricate the engines of a supersonic jet or to make the cellulose acetate for films, but plants are still just as important to our lives as they were to our early ancestors on the savannas of Africa. The diversity of plant species of the world is so useful to us that it is surprising that contemporary society is allowing any plant species to go extinct before its full potential for use has been researched.

We have here a fascinating account of the many ways in which plant products surround us in our daily lives. Although we tend to think of the dependence of indigenous peoples upon the plants of the forest that surrounds them, we often forget that the soap with which we wash, the breakfast we eat, many components of the motor car we drive, clothes, medicines, stimulant drinks such as coffee or tea and building materials are all from plants. This book is interesting because it describes and illustrates both examples of indigenous tribal uses of plants and the plant products that enter our daily life in a modern city: for example, from the body paints of the Amazon Indian to modern cosmetics, or from early Egyptian papyrus of over five thousand years ago to today's pulp mills that produce the 200 million tonnes of paper that we consume each year.

A great amount of botanical and anthropological data is presented here in a highly readable form. Any reader is sure to learn new facts about products that we take for granted. The progressive development of humankind has been based on the availability of a great diversity of plants. It is salutory to be reminded here of our reliance on green plants. I hope that a side-product of this most interesting book will be our realization that we must protect the future of the plant species upon which life depends.

Ghillean T Prance
ROYAL BOTANIC GARDENS, KEW

Starting the day

We spend most of our lives with our bodies covered with plants! In the form of oils and extracts, gums, waxes and fibres, the vast proportion of the things we use to wash and dress ourselves each day involves them either directly or indirectly. By the time we have opened the door in the morning we will almost certainly have used dozens of different plants.

WASHING WITH PLANTS

Most of us will use soap in one form or another to start the day. In Britain alone over £85 million was spent on toilet soaps in 1988, and world production was estimated at over 8 million tonnes. The base for British soap manufacture has traditionally been tallow, a coarse hard fat obtained chiefly from cattle and sheep. Now, however, plant oils, particularly from the African oil palm and coconut palm, are extensively used in the industry, to produce the vast array of soaps that are on sale in supermarkets and drug stores around the world.

Pockets full of posies

Soap perfumes may be derived to a large extent from plants. Many aromachemicals are now synthesized from turpentine taken from various species of pine tree, adding to the range of natural fragrances produced by individual plants. Indeed, it was these natural aromas that led people to use plants to banish body odours and unpleasant smells in general long before the invention of the bar of soap.

From the Dark Ages onward Britain and the rest of Europe relied heavily on the scent from aromatic herbs to alleviate the generally malodorous conditions that existed. Since hygiene itself seems to have been almost non-existent at this time, floors were strewn with aromatic plants both to mask bad smells and to deter fleas, lice and vermin. Such plants included various mints, sweet rush, sweet woodruff, germander, camomile and hyssop, which released their fragrance as they were trodden on. Posies of fragrant flowers were also carried in the belief that they would ward off disease.

Long before this use of herbs in Europe however, the ancient Egyptians were using the oils from plants and fragrant flowers as part of their ablutions. The Romans threw lavender into their baths, not only to scent the water but to act as a disinfectant. In fact the scientific name for lavender 'Lavandula' is derived from the Latin 'lavare' meaning to wash.

Ring a ring of roses
a pocket full of posies. .
runs the nursery rhyme. Though we no longer carry posies of fragrant flowers to ward off disease, their use in our homes – to sweeten the air or freshen clothes – continues often in the form of pot-pourri.

The first soaps from plants

The use of soap in tablet form for washing did not become widespread in Europe until the mid-nineteenth century, but its manufacture has a long and interesting history. Though their role has changed somewhat, plants have been essential to the soap-making process since the earliest times.

The Egyptians are known to have invented a sort of natural washing soda by mixing wood ash and animal fat, and by AD 70 the Romans were making soap by burning goat tallow with cauterized beech ash. A soap maker's shop was in fact preserved beneath the lava and ash of the eruption that smothered Pompeii in AD 79.

One of England's earliest centres of soap-making appears to have been in Bristol during the twelfth century, but by the fourteenth century the famous Castile soap was being exported from Spain, manufactured using olive-oil instead of goat tallow. Soap-making became established in Marseilles, Venice and later Savona, where the most sophisticated soaps incorporating exotic perfumes and aromatic powders were made.

During the seventeenth century, English soaps, which were soft and often made from fish oil (still used today in some), were still a luxury. Less wealthy but still fortunate households used wash-balls which they could make themselves from a mixture of plant oils, shavings of Castile soap, distilled water and various herbs and spices. Ground almonds, raisins, brown breadcrumbs and honey were also sometimes added.

By this time the ash from burnt beech wood, which had formed the alkali for many soaps had been replaced by that obtained by burning sea plants, especially the spiky leaved saltwort (*Salsola kali*), which grew commonly on sea-shores and salt marshes. This ash, known as barilla, was widely used by soap and glass makers until 1787, when it was discovered that alkalis could be made from common salt. The discovery now gave the soap-maker access to an unlimited supply of one of the basic raw materials. However, it was not until the Industrial Revolution that increased wages together with mass production methods gave ordinary working people the chance to buy a bar of soap.

Soap manufacture today

Two palms provide us with the great bulk of the plant oil used in soap production: the African oil palm (*Elaeis guineensis*) and the coconut palm (*Cocos nucifera*). Several other plant oils can be utilized instead of or as well as these, among them soyabean, cottonseed and groundnut. Which oils are chosen will depend to a large extent on where the soap is made and the price and availability of each one.

The African oil palm is a particularly valuable tree since its fruits yield two different types of oil (palm kernel oil from the seeds and palm oil from the fibrous pulp which surrounds them) for both food and non-food uses. This palm is now one of the world's most important providers of edible and soap-making oils, yielding more oil per year per hectare than can be obtained from any other vegetable or animal source.

Whilst much of the oil from the African oil palm is used for food, chiefly the production of margarine, soap and cosmetics manufacture are the most important commercial uses of coconut oil. This is extracted from the edible white flesh of the coconut, which is first dried either in the sun or artificially to form copra, the tree's most valuable product.

Plant oils are now extensively used in soap manufacture, especially to make the finer toilet soaps. They have two main advantages over tallow. Firstly they produce soft, clear-textured soaps which lather well and are less prone to cracking; secondly the price is very favourable compared with that of animal fats.

Though olive-oil from the Mediterranean has been used in soap manufacture for centuries, the large scale move from tallow to tropical plant oils (African oil palm, coconut, peanut etc.)

COCONUT PALM

Cocos nucifera

The coconut palm can be found growing along almost all coasts of the tropics where the rainfall is greater than 1.5 metres per year. Tolerant of salty, sandy soils it will grow on strips of land where most other crops could not survive. Mature trees can reach a height of 24 metres, developing slender, often curving trunks, with a distinctive feathery crown of leaves at the top.

The fruit of the coconut palm — from which coconuts are extracted — appear from about the sixth year onward, and may be produced until the tree is about 80 years old. Green at first, they turn yellow as they ripen. The tall, branchless trunk of the palm can make harvesting very difficult, but skilled bare-foot climbers often use ropes to help them ascend the trees and in some South East Asian countries monkeys are trained to do the job, throwing the fruits to the ground as they reach them. Alternatively, the fruits may be left to drop of their own accord or are cut from the trees by knives attached to long bamboo poles.

All parts of the fruit are useful. The thick layer of fibrous husk that lies beneath the outer skin for example can be combed out and sold as coir, an important material used for making ropes and matting. The object that we call the coconut lies within this husk and is actually the nut of the fruit. Inside the nut's shell is the coconut 'meat', the edible, white, fleshy layer, and within this, a cavity partially filled with 'coconut milk', a watery liquid containing sugars, that is gradually absorbed by the fruit as it ripens.

The coconut palm has a great number of uses, and has become a vital component of many local economies. Its trunks provide building timbers and the leaves material for thatch. Wood and leaves are also used as fuel, whilst the fruit provides food, drink and fibres. The most important commercial product of the coconut palm, however, is copra, the dried coconut flesh. The Philippines currently export the most copra, whilst Indonesia, Malaysia, India and New Guinea are also major exporters. Though produced in small quantities copra and coir are particularly important to many Pacific Island communities.

Fruits from the African oil palm

Women from Cameroon crushing oil from the fruits

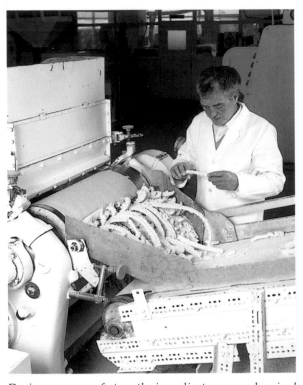

During soap manufacture the ingredients are evenly mixed by milling the mixture three times. This process also stops the soap from cracking and going soggy. Almost all soaps produced today, including those illustrated here, involve plant oils in their production.

did not come about until the beginning of the twentieth century, when the world experienced an acute shortage of animal fats. At this time a means of hardening or hydrogenating plant oils previously considered too soft or liquid for use in soap manufacture was invented in Germany, and the new vegetable sources were a great success.

In very simple terms, soap is produced by reacting the tallow and/or plant oils with alkalis — commonly caustic soda or caustic potash — in a process known as saponification. Plant oils are made up of glycerol and fatty acids in combination and saponification splits these to yield sodium or potassium salts of the acids plus glycerol. Molecules of these salts, which are technically soap, comprise two parts — 'water-loving' sodium or potassium at one end and 'water-hating' fatty acid at the other. This duality enables the water-hating (hydrophobic) tails to attach themselves to greasy dirt particles, whilst the water-loving (hydrophilic) heads suspend the dirt in the water, allowing it to be carried away.

Most toilet soaps are still made from a mixture of animal fats and vegetable oils which are blended together before saponification, and the characteristics of the finished product will

ALTERNATIVE SOAPS

Saponin is a generic term applied to a range of organic compounds which produce soapy, frothy solutions. Many plant species contain saponins and will produce appreciable amounts of cleaning lather when rubbed or boiled in water. In Europe the plant most commonly used in the past was soapwort (*Saponaria officinalis*) which was also known as fuller's herb, latherwort, crow soap and soap root.

In Sudan and Chad, a plant oil used by the Egyptians some 4000 years ago is still in use as a type of soap today. It is produced by crushing the bark and seeds of a small spiny tree, *Balanites aegyptiaca* which yields a yellow oil, rich in saponins.

In Arabia and Somaliland, some desert peoples clean their bodies by crouching over a smouldering pot of charcoal and aromatic herbs, gums and spices, their robes spread around them to catch the smoke and promote heavy sweating. The inhalation and fumigation of the body with smoke from the burning of fragrant plant materials is in fact one of the oldest uses of plants by man — thought to both cleanse and regulate the body and balance any 'misalignment' of the body fluids or 'humours' which are believed to determine temperament.

SOAPWORT

Saponaria officinalis

This herbaceous perennial, which will grow to a height of 1.5 metres, was once widely used for washing. When boiled in water the leaves and root produce a green soapy solution, which is still occasionally used today for the cleaning of very

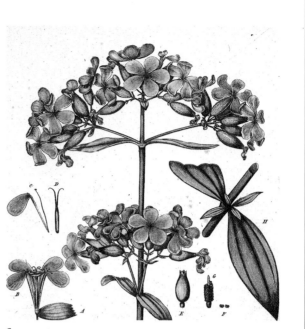

Soapwort

old and delicate fabrics such as tapestries. One early herbal describes the 'great scouring qualities' of the leaves, which 'yield out of themselves a certain juice when they are bruised which scoureth almost as well as sope'.

The increased use of animal fats as a base for soap, however, meant that soapwort became outmoded in England, and from the early seventeenth century was seldom mentioned in British herbals except as a medicine 'for the French pox'.

Today, saponins derived from other sources find a major use in the food industry. They are particularly useful in the manufacture of beer and lager, helping to produce and maintain a good head of foam. (*Alternative soaps cont. p.6*).

depend largely on the proportion of one to the other. The most common blend comprises some 80 per cent tallow to 20 per cent plant or 'nut' oils, but a more luxurious toilet soap will have up to 40 per cent plant oils and 60 per cent tallow. Good quality soaps require specific blends of fatty acids for the right balance of solubility, detergency and lather. 'Superfatted' soaps are rich in fatty acids and these stabilize the lather giving a creamy texture.

Used on their own, soaps made from plant oils would dissolve away too quickly and lather too much, so hard, white tallow is added to give firmness. In the future, however, as tallow becomes more expensive, it is expected that the soap industry will change to the use of palm kernel oil which is chemically very similar to tallow, to perform this function.

The fats and oils used to make soap undergo a series of refining procedures before saponification, after which ingredients such as colour and preservatives (mainly made from inorganic materials) and perfume will be added. Glycerine, itself a by-product of soap manufacture, washed out of the mixture by boiling with brine, is sometimes re-introduced to the soap to enhance its softness and appeal.

Balanites aegyptiaca

Atuqsara

Quillaja

ALTERNATIVE SOAPS

Balanites aegyptiaca

Indigenous to the Nile Valley, this small tree has great potential for the people of the Sahel region in Africa. It has been estimated that every inhabitant could collect up to 450 kilogrammes of fruit and leaves per year from local trees, while in Sudan a yield of about 400 000 tonnes of fruit has been suggested. The fruit can be processed to give not only saponins (for commercial use), but edible oils, a substitute for peanuts and animal feed. Since the use of *Balanites aegyptiaca* as food and fodder is limited by the presence of saponins however, a saponin-free or very low content form needs to be found and selected for cultivation before the full potential of this species is realized.

Perhaps the most important use of *B. aegyptiaca* is medicinal. The root, bark, fruit-pulp and seeds have all been found to be lethal to fish and fresh water snails which act as an intermediate host for bilharzia. The presence of an extract of the tree in water can also kill minute free-living forms of the bilharzia parasite, which affects millions of people in tropical countries each year.

ATUQSARA

Phytolacca bogotensis

In the highland regions of southern Ecuador, local Quichua-speakers with little money to buy western goods continue to gather the fleshy, fruiting heads of atuqsara (literally 'wolf's maize'). The plump red berries containing saponins are crushed in water and used to wash the hair as well as clothes.

QUILLAJA (the soap tree)

Quillaja saponaria

Another South American plant — this time a tree native to Peru and Chile which grows to around 18 metres in height — yields saponin in much greater quantities than atuqsara. It is extracted from the tree's bark and after purification is made available commercially in Europe in either liquid or solid form for use in shampoos, foam baths and other toiletries.

Besides giving a formula for a home made shampoo based on quillaja, also known as soap bark, Mrs Grieve (in her *Modern Herbal*, 1931), lists some of the more surprising attributes of saponin, which 'when applied locally, is a powerful irritant, local anaesthetic and muscular poison'. Used in very small and diluted quantities however, saponin is quite safe.

SHAMPOO

Just as most households never lack a bar of soap — or at least try not to — shampoo, in some form or other, is rarely absent from our bathrooms.

More time and money is probably spent caring for our hair than any other part of the body! Once again, plants are indispensable. They are often largely responsible for the cleansing properties of the shampoo itself and add thickness as well as fragrance to the base.

A myriad of plant names entice us to buy shampoos, advertising extracts and oils that claim to beautify and tame our hair — from lemon to nasturtium. Though shampoo is a more recent introduction than the bar of soap, aromatic plants and flowers have been used since the earliest times to beautify the hair. Aristocratic Chinese ladies, like many Eastern women are said to have oiled their hair and bound into it strong-scented flowers such as frangipani and

JOJOBA

Simmondsia chinensis

Jojoba is one plant that has gained enormous popularity in recent years, and is likely to appear in shampoos in increasing quantities. The oil is pressed from the fruit of *Simmondsia chinensis*, a low-spreading bush that grows wild in the Sonoran Desert of Mexico and parts of the south-western USA. Its seeds contain up to 60 per cent of a light yellow, odourless liquid wax — a wax which proved to be an almost perfect substitute for the sperm whale oil needed for the production of ballistic missiles. The millions of dollars invested in the future of jojoba oil were essentially to provide a natural raw material for military use, but shampoo, hair conditioners and a range of cosmetics have, in taking advantage of the plant's exceptional qualities, promoted its popularity.

The remarkable qualities of the oil pressed from the inconspicuous fruits of *Simmondsia chinensis* are a reflection of its unusual composition. Containing large molecules made up of long carbon chains, the oil would under normal circumstances be a wax. A structural hindrance to solidification, however, means that the oil remains liquid and so is technically termed a liquid wax.

Apart from its use in shampoos and cosmetics, jojoba oil has been used in the manufacture of engine lubricants, printing inks and varnishes. In solid form it can also be used as a polish and to protect many different surfaces.

jasmine, which perfumed it long after the flowers had died. Today, Tahitian women still use a pommade made of coconut oil which has been scented with sandalwood or toromeo root, to add a fragrant shine.

The main ingredients of today's shampoos are water (up to 80 per cent), detergents and perfume, to which are added thickeners, lather-boosters, colouring agents and preservatives. Since soap, as we have just described it, would leave a scum on the hair and also dull it, the cleansing agent in most modern shampoos is usually a highly soluble synthetic or non-soap detergent. Although the starting materials of this detergent, fatty alcohols, may be derived from petrochemicals or tallow, they are often made from natural plant oils. Coconut and palm oils are likely to be used. To improve detergency and the stability of the foam, 'secondary surfactants' as they are known in the trade are also present in most shampoos. Chemically these are fatty acid derivatives, which are again mostly taken from palm and coconut oils.

The smell of our shampoo is an important part of its appeal. Since, for example, a strong mental association exists between lemon juice and the removal of grease, shampoos formulated for greasy hair are often given a 'lemon' or other 'citrus' smell. Whilst many of these fragrances are chemically synthesized, the starting materials used to make them are often plant-derived.

The herbal extracts and oils that many shampoos boast and which are alleged to enhance the performance of the product in some way, are as they claim, extracted from the plant in question. Rosemary, nettle, horsechestnut, or camomile may all be present in extract form along with oils such as jojoba, apricot or peach nut, but often in only very small quantities — sometimes less than 2 per cent of the total volume.

Thickeners

Whilst the colouring and preservatives in most shampoos are almost entirely synthesized from minerals or mineral oils, some very important components of shampoos and liquid soaps — the thickeners or viscosifiers — are usually derived from plant cellulose, the chief component of the cell membrane of all plants.

Plant cellulose is the raw material behind a world-wide, multi-million pound industry, which produces a range of thickeners and suspending agents essential now not only to a large number of cosmetics and toiletries, but also to the manufacture of many convenience foods, and the paint, printing and detergent industries. Many hundreds of thousands of tonnes of cellulose are used each year to help make products as diverse as fish fingers and hair mousse.

As an ingredient of shampoo, cellulose which has been specially treated to produce a substance which will dissolve in water (a water-soluble polymer) is added in the form of colourless granules. Though present in relatively small quantities in relation to the total weight of the product, polymers can be made to give a range of consistencies, from very thin liquids to thick gels which help bind all the ingredients together. The polymers added to shampoos do not affect its clarity but help make the foam produced heavier, creamier and more stable. They are also said to improve the manageability of shampooed hair.

Bath oils, skin fresheners and cleansers, aftershave lotions, deodorants and anti-perspirants may all contain cellulose as a thickener. In colognes and perfumes, water-soluble polymers often act as binders, holding the odour components to the skin to lengthen the time that they will be effective.

To produce these polymers (often either methyl cellulose or sodium carboxymethyl cellulose (CMC)), plant cellulose in the form of wood pulp and/or cotton linters is first reacted with an alkali and other chemicals and then purified, dried and powdered. Natural cellulose does not dissolve in water, but these changes to its chemical structure will make its derivatives soluble in either hot or cold water or alcohol, according to the formulation. Each gramme of polymer, which will be classified as either high or low viscosity, will contain about 0.75 gramme of plant cellulose.

Whilst wood pulp generally produces the lower viscosity polymers, cotton linters (the short stiff fibres which remain on the cotton seed after the removal of the longer cotton fibres), are almost pure cellulose and therefore produce a higher viscosity. Some British manufacturers use South African wood pulp, often containing up to 85 per cent eucalyptus, with wattle (*Acacia* spp.) as the next biggest ingredient. European hardwoods and softwoods such as spruce and beech from Scandinavia and Finland are also used but tend to be more expensive. Generally speaking, North America is the largest overall producer of pulp for cellulose derivatives. Large quantities of mixed American hardwoods, including various species of oak, as well as western hemlock from the West Coast are converted into water-soluble polymers each year. North America also supplies most of the cotton linters used.

Hair conditioners

Unlike shampoos, the primary active ingredients in most modern hair conditioners are extracted from tallow rather than plant oils. They do, however, contain water-soluble polymers made from cellulose, as do many hair gels and mousses used for grooming and styling, which leave a very thin, clear, flexible and non-sticky film on the hair shaft after drying.

Selected plant oils such as galbanum, distilled from the aromatic gum resin produced by various species of *Ferula* which are native to Persia, are present in some conditioners in small quantities, as well as protein, said to give a strong protective coating to the hair shaft and add body and lustre. Though the protein may be derived from animal fat, it is often made from soyabean oil, one of the cheapest and most widely used oils in the cosmetic industry.

A great number and variety of plant materials — and some highly unpleasant animal ingredients — have been used throughout the ages to strengthen, thicken and perfume the hair. Besides relatively innocuous gums and woods, all kinds of alcohol and the ashes of burnt bees and goats' dung were once used. Puppys' fat and bears' grease have also been recommended in the past. Our modern use of lanolin, the fat extracted from sheep's wool, in conditioners and shampoos, seems a fairly civilized alternative! Most famous in Victorian times for the well-groomed gentleman's hair was macassar oil. It is thought to have been based

THE LAC TREE

Schleichera oleosa

In India the lac tree is grown for its seeds, which contain the edible fat thought to be the original source of macassar oil. This fat is still used today for hair oils and is burnt as a source of illumination. The fruit pulp can be eaten raw, the unripe fruits are consumed as pickles and the young leaves serve as a vegetable.

Small boats and other items are made from the hard red timber and the bark can be used for tanning.

on the oil obtained from the seeds of the lac tree (*Schleichera oleosa*) also known as the Ceylon oak, kussum tree and Malay lac tree which is native to India and parts of Asia. Though this oil is no longer used in Britain, rectangular cloth coverings for the backs of chairs, originally designed to protect them from macassar oil, and thus called 'antimacassars', are still found in many homes.

BATHING WITH PLANTS

A range of plant oils and extracts are also available at bath-time to soothe and relax us. As in the case of soaps and shampoos, coconut and palm oil often provide the cleansing and foaming bases of bath and shower gels, and in the form of fatty alcohols act as solubilizers for numerous perfumed bath oil constituents. Citric acid made from sugar cane and sugar beet may be added to help the product match the pH value of the skin's outer surface, and molasses is again sometimes the source of substances used to keep the skin supple and moisturized. But perhaps the most useful plant for bath products is the castor oil tree.

Very few oils disperse completely in water, but castor oil, sometimes sold as Turkey Red oil, from the plant *Ricinus communis*, not only disperses but does not mark the bath! Native to India and tropical Africa, but now widely distributed in many tropical and temperate regions, the plant was known to the Egyptians, and its oil, pressed from the seeds, was used as an unguent base as well as a fuel for lamps. Today, castor oil is an important ingredient of many bath preparations, helping to soften and lubricate the skin.

The castor oil tree (Ricinus communis) *is a native of the tropics, where it grows up to 4 metres high. The pale yellow oil extracted from the highly toxic seeds is well-known for its laxative properties and is also used for the manufacture of resins, plastics, lubricants, and in bath oils.*

An advertisement for lavender from Mitcham, Surrey c. 1890

Harvesting at the Norfolk lavender farm

Of all the plant oils and extracts that are or could be used for the bath, lavender is one of the oldest and best known. The centre of a large and very valuable industry today, lavender oil or extract has been popular as a bath ingredient since the days of the Romans and continues to make its way into the most popularly used toiletries and scents. At one time Britain was famous for its lavender fields and Mitcham in Surrey was the centre of the industry until the early twentieth century. Lavender is still grown commercially in Norfolk, but most of Europe's supply now comes from Italy and France. Provence in particular is favoured by those with a nose for the finest aromas.

Fields of lavender in Provence, France

LAVENDER

Lavandula spp.

A small shrub commonly grown in many European gardens, lavender, with its distinctive narrow, grey-blue leaves and spikes of soft purple flowers, is native to the Mediterranean area.

Some eight species are recognized but *Lavandula vera* and *L. angustifolia* are the main two grown commercially, both for their flowers, which retain their fragrance when dried, and their aromatic oils much used in perfumery.

Lavender plants are very sensitive to heat, light and different soils, and the oil produced will reflect growing conditions. Whilst the lavender currently grown in Norfolk is said to be very sweet, that from Provence is widely used in aromatherapy and is said to help balance the mental and physical energies. Long used in medicine, lavender oil is also valued as a cicatrisant, helping to heal scars.

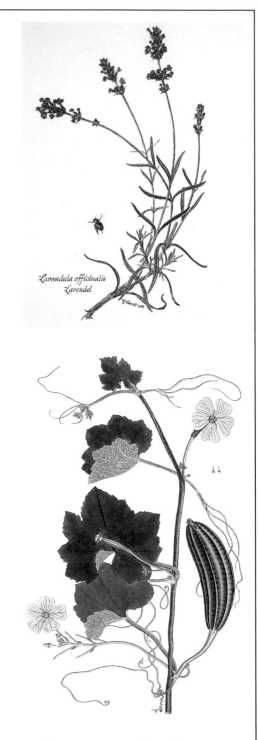

Lavandula officinalis
Lavendel

SPONGE GOURD

Luffa aegyptiaca, L. acutangula

That familiar bath accessory, the loofah, is the bleached skeleton of a fruit produced by a member of the gourd family. Two species of *Luffa* are widely cultivated, but *L. aegyptiaca* is the one most widely used in commerce. The genus name and our own term loofah are derived from an Arabian name for the plant, which is thought to have originated somewhere in Asia.

L. aegyptiaca has deeply lobed leaves and large bright yellow flowers which open in the morning sun. Its fruits are cylindrical in shape and smooth, growing to some 0.5 metres in length. Though green when young, they become a yellow or tan colour when mature.

In Japan, which has been growing the gourds commercially since the 1890s, the vines are trained over trellises to give maximum exposure to the sun, and some 20 to 25 fruits are allowed to mature per vine. To extract the loofah 'sponges' the fruits are placed in tanks of running water until the outer walls disintegrate, and further washing removes any seeds or pulp still clinging to the skeletal fibres. The 'sponges' are then left in the sun to dry and bleach — a process which is completed with hydrogen peroxide — before they are graded and packed for export.

Apart from their use in the bath sponge gourd skeletons have been used for cleaning cars, glass-ware and kitchen utensils. Until World War II, the USA imported large quantities from Japan, some for cleaning purposes, but 60 per cent were utilized as filters for steam engines and diesel motors. With the outbreak of war, production was started in South America, but the loofahs from Japan are generally acknowledged to be of superior quality.

In South East Asia the whole sponge gourd is eaten as a vegetable when young and tender.

A fragrant oil is to be found in all parts of the lavender shrub, but the essential oil of lavender is extracted only from the flowers and flower stalks, usually by steam distillation. Lavender is often sold dried as an ingredient of pot-pourri and in sachets for scenting clothes and bath-water. Besides scenting the water, it is said to stimulate the action of the skin's pores, relax muscles and soothe the joints. As the Romans discovered, lavender's antiseptic properties make it a useful addition to the bath.

There is one plant though — or rather plant fruit — that many of us actually use as an instrument to wash ourselves. It is commonly found residing in bath racks all over Britain! The loofah as it is widely known may look like the dried form of some strange sea sponge, but it is actually the bleached skeleton of a tropical fruit, produced by the sponge gourd, a large climbing plant native to Asia. Removal of the papery rind of the mature fruit reveals the familiar fibrous network which we use to scrub ourselves. In Japan, the main commercial supplier of these gourds, the juice extracted from the plants' stems is highly prized for annointing and softening the skin. It has also been used for respiratory complaints and as an ingredient of toilet water.

Talcum powder

Even after we step out of the bath or shower and sprinkle our bodies with talcum powder we may still not be free of plants. Though the traditional ingredient of talc (from the Persian 'talq', a soft greasy powder first introduced into Europe in the sixteenth century) is powdered hydrated magnesium silicate, or mica, the term talcum powder is used rather loosely today and a number of other substances may be included. Apart from calcium carbonate (precipitated chalk) now commonly used, talcum powder may contain finely powdered maize or rice starch. Following the fears that the constant use of mica may be cancer-inducing, a movement away from this traditional ingredient in favour of vegetable powders is now underway.

Swiss Herbal Toothpaste with Fluoride

50 ml 1.8 FL OZ

TOOTHPASTE — a brush with plants!

Having bathed or showered with the help of plants, how do you fancy brushing your teeth with wood pulp?

If you use a standard toothpaste you will probably be using it twice a day already! Carboxymethyl cellulose or cellulose gum, derived as we have seen from wood pulp and cotton linters, is used in greater quantities in toothpaste manufacture than in any other cosmetic or toiletry item.

Though it may only form some 1 or 2 per cent of the total weight of the paste (over half of which is comprised of abrasives) it holds all the other ingredients together in a free-flowing gel that will dilute easily in water. Cellulose also suspends colour, helps the toothpaste keep its shape when squeezed onto the brush — yet disperse evenly in the mouth — and enables it to be rinsed free of the toothbrush after brushing.

This is not the only plant-derived ingredient in toothpaste. Humectants, which prevent the paste (usually about 20 per cent water) from drying out and help preserve its flavour, are traditionally made from glycerine, which may be derived from plant oils, or sorbitol made from wheat or maize starch. Humectants will often comprise about one quarter of the

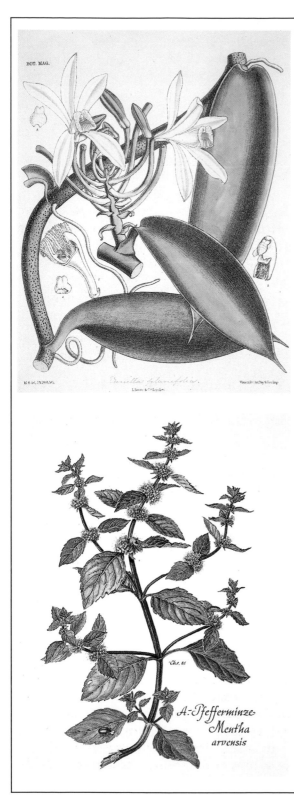

VANILLA

Vanilla planifolia

The vanilla plant is a tropical orchid, native to South and Central America. The seed pods take nine months to mature and, after picking, several more to sweat and dry out (a stage known as fermentation), during which time the distinctive scent and flavour compound, vanillin, is formed.

Used very widely in perfumes and confectionery, as well as in toothpaste, vanillin is extracted by continuously percolating an ethanol mixture over the chopped, fermented pods, thereby dissolving it out. The very labour-intensive production, as well as its scarcity relative to demand, has made vanilla the second most expensive spice in the world after saffron. Vanillin is now synthesized for commercial use on a large scale, sometimes from wood pulp.

MINT

The flavour and aroma of mint plants comes from aromatic essential oils which are secreted by glands in the leaves and stems. As well as their uses as flavourings for toothpaste, oil of peppermint, distilled from *Mentha piperata* and *M. arvensis*, is a likely ingredient of mint flavoured sweets, whilst *M. spicata* is commonly made into a jelly or sauce and served with roast lamb.

Mentha arvensis

toothpaste's weight. The foaming agents or detergent (such as sodium lauryl sulphate) will often have plant oils, especially coconut or palm, as their base, whilst alginates from seaweeds may be added as thickeners.

If one set of plants is widely associated with toothpaste it must be the mints. Though some synthetics are used, the minty flavour of most toothpastes is derived from natural mint oils.

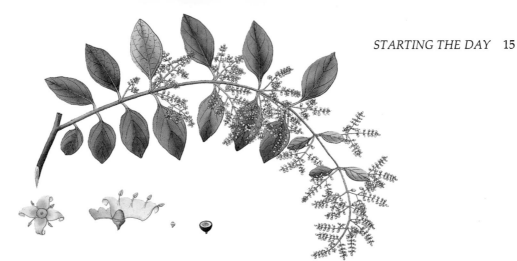

The toothbrush tree (Salvadora persica)

Four different species of mint are commonly used: *Mentha piperata*, better known as peppermint; *Mentha arvensis*, sometimes called peppermint by mistake since both contain menthol; and two varieties of spearmint, *Mentha spicata* and *Mentha cardiaca*. Much of the peppermint and spearmint oil used in Britain comes from producers in the USA, especially from areas such as Indiana, Wisconsin and Oregon. *Mentha arvensis* and a certain amount of spearmint oil also come from China and India. Brazil was once a major source of mint oils, but a shift towards greater coffee and soyabean production has meant a reduction in this trade. Other important mint producers include the USSR, Italy, Japan, Thailand, Paraguay, Taiwan, the UK, and Germany.

While it accounts for over half of the 500 000 tonnes or so of flavouring materials used in toothpastes in the Western world each year, mint is not the only natural flavouring from plants. Vanilla extract is very common in many toothpastes, whilst some are flavoured with aniseed, fennel or cloves as well as citrus oils.

Before leaving the subject of toothpaste, it is worthwhile remembering that your toothbrush may also owe its existence to plants! Many moulded plastic toothbrushes are made from cellulose acetate — itself derived from wood pulp reacted with various chemicals.

Wood for teeth!

Not everybody cleans their teeth with a toothbrush and toothpaste. In parts of East Africa, chewing sticks are used to keep teeth clean and help prevent mouth infection. At least two of the *Diospyros* species involved (*D. usambarensis* and *D. whyteana*) have been shown to contain potent anti-fungal compounds. Also in Africa, as well as parts of India and the Middle East, the woody stems of *Salvadora persica*, the toothbrush tree, have been used for centuries to help clean teeth and gums. When chewed, the stems release juices which appear to have a protective anti-bacterial activity, helping to offset decay and gum disease. An extract from the toothbrush tree is now incorporated into some commercial toothpastes sold in Britain.

If, after all our chewing or brushing we still need dental care in the shape of false teeth, plants may still be helping us. Gum karaya exuded from large trees of the genus *Sterculia*, which occur widely in India, Sri Lanka and many parts of Africa, is an important dental fixative in the West. Its special adhesive properties as well as its resistance to enzymatic and bacterial breakdown have led to its wide use in dentistry, helping to keep false teeth, veneers, plates and crowns in place.

COSMETIC CREAMS AND LOTIONS

Plant derived materials offer far more to the advertising or marketing manager, as synthetics and animal source materials lack glamour.*

This is the view of one of Britain's leading suppliers of plant extracts and raw materials to the cosmetic industry. Under the marketing manager's directive, plant-derived materials are set to become even more important in the race to win customers from a rushed, polluted and, ironically, plant-threatening world. The idea of good health and correspondingly beautiful skin, associated with life in the country or simply certain plants or flowers, can be sold to us quite simply with the help of a pretty label and a name.

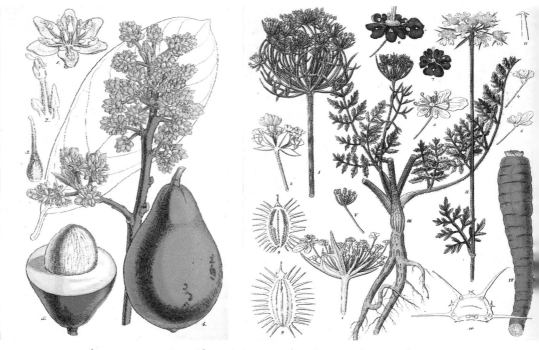

Oils extracted from macadamia nuts, avocados and carrots are all used in the preparation of cosmetics

Macadamia tree

But a general movement away from expensive if enticing packaging and complex 'secret formulae' on the part of at least one well-known manufacturer[†], and a desire on the part of the public to buy products that are, 'environmentally sound', has led to a new emphasis on the active role and importance of plants as beauty aids.

Despite the current trend to advertise if not list the herbal origins of creams and lotions for smoothing, protecting, enlivening and 'de-ageing' our skin, many of the basic ingredients — plant oils, extracts, starches and gums — as well as disparate items such as eggs, beeswax, honey and lanolin, have in fact been used for centuries, but often in much simpler ways.

The first recipe for cold cream, for example, as recorded by the Greek physician and teacher Galen in the second century AD, mixed water with beeswax and oil of roses. Both of these are used today for various cold cream formulations, though the very expensive oil or attar of roses is often replaced with cheaper vegetable oils.

Plant oils certainly have a very ancient pedigree where beauty treatments are concerned. Olive-oil, perfumed with saffron and other aromatic plants was used by the Egyptians to keep their skin supple, and later the Romans, who had learned about cosmetics from Egypt, used

* D.Mitchell & M.James, (Chesham Chemicals), in *Manufacturing Chemist*, Oct. 1987, p57.
† The Body Shop International PLC.

ALMONDS

Prunus dulcis var. *dulcis*; *P. dulcis* var. *amara*

Almond oil is one of the most important of the fine cosmetic oils, and is pressed from the kernels of the ripe fruit of two varieties of almond. *Prunus dulcis* var. *amara*, the bitter almond, is the main source of the almond oil which is used for flavouring as well as cosmetics but the oil from the sweet almond (*Prunus dulcis* var. *dulcis*), which is grown for its delicious nuts, is also used in skin preparations. The oils are light and make very good emulsifiers and emollients, softening and smoothing the skin. Though native to South West Asia almond trees are grown extensively in Mediterranean countries, especially Italy and Spain and are cultivated on a large scale in California, China and Iran.

The distinctive almond smell and taste is due to the presence of prussic acid in the almond kernels.

pumice stones to rub their bodies with this much-prized lubricant. Though olive-oil is still found as a cosmetic base today, other cheaper oils often replace it. Soyabean is the most widely employed in cosmetic preparations because of its price. Peanut oil is also popular, along with sunflower, castor, and the ubiquitous coconut and palm oils. Whilst these oils, after refining and deodorizing to give better colour, odour and stability, may replace the mineral oils that now commonly form the base of many modern creams and lotions, a wider use is the production of a range of chemical derivatives.

Many vegetable oils are triglycerides, that is they have three fatty acids attached to their glycerol molecule. The fatty acids most commonly occurring in vegetable oils are palmitic, stearic and oleic, and these can be separated by a process of fractionation for individual use in different formulations or can be further converted into a number of other compounds. Reacted with fatty alcohols for example, from both plant and non-plant sources, they form esters, important in the manufacture of perfumes and flavourings.

Fatty alcohols obtained from plant oils such as soyabean, linseed, rapeseed, peanut, palm and coconut, are often added to emulsions to help adjust viscosity. The esters of these fatty alcohols also act as emulsifiers, combining substances that cannot normally be mixed, such as water and oil, and evenly distributing solid particles in a liquid medium.

A large number of other, more expensive plant oils can be added to cosmetic preparations for their own particular characteristics. Often expressed from plant seeds or nuts, some of the most popularly used include almond, avocado, apricot and peach nut. Even the humble carrot may be pressed — quite literally! — into service yielding a yellow oil that is said to soothe chapped skin and combat dry hair.

With our desire for an ever more youthful appearance, a large proportion of the skin preparations we buy are sold to help revitalize or moisturize the skin. Whilst most plant oils will help soften it, macadamia nut oil (extracted from the fruit of *Macadamia tetraphylla* and *M. integrifolia* trees native to north-eastern Australia), is of special interest. With a fat content of over 70 per cent, the nuts contain uniquely high levels of palmitoleic acid, a component of the skin's epidermal fluids, said to be responsible for its softness, suppleness and water-retaining capacity. The amount of these lipids and palmitic acids decreases in females rapidly after the

The cocoa pod

age of twenty, and so this nut oil has been suggested as a suitable component of creams intended to improve the appearance of ageing skin.

Cocoa butter, a cream-coloured waxy solid pressed from the shells or husks of the seeds contained within the cocoa pod is another useful substance, often employed as a lubricant and moisturizer. Melting at body temperature, it gives the skin a silky finish. Artificial cocoa butter used for the same purpose is often made from palm oil after a process of refining and fractionation.

Also derived from plants and acting as a humectant, especially in cosmetic creams, is glycerine which is produced by splitting triglyceride plant oils. Lecithin, a by-product of soyabean oil processing, is another humectant and emollient. It too is added to cosmetic creams, to act as a foam stabilizer and anti-oxidant, stopping the plant oils from turning rancid. Lecithin can be purified and hydrogenated to make 'liposomes' — sold as the fashionable de-ageing ingredient of some face and body creams. Because of their resemblance to human skin cells they are said to enable active ingredients held together in emulsion form to penetrate the skin and slow down the ageing process.

Not to be forgotten, and playing an indispensable role in skin preparations and most other cosmetics, are the thickeners. As we have already seen, plant cellulose provides the raw material used to make an important range of polymers which have this function, and it is frequently present in the beauty treatments we buy.

Other naturally occurring plant materials are also widely used. They include gums from flowering plants and seaweeds, and starch, often from potatoes. The plant gums most often used in the production of cosmetics and in a large number of other industries, from food processing to oil drilling, include guar, locust bean, gum tragacanth and gum arabic. All these can be used to thicken and stabilize creams and lotions, as well as shampoos, adding spreading properties to the product and a smooth, protective coating to the skin or hair.

Gum arabic for example, the most widely used of all the natural gums in industry as a whole, makes an excellent adhesive in face masks, helping to keep the liquified ingredients from dripping down the face! While the true gums — arabic and tragacanth in this case — come from woody plant tissue such as roots or branches which are slashed or punctured to cause the gum to be exuded, locust bean and guar gums, which are similar substances are produced from the seeds of leguminous plants.

As well as these terrestrial plant gums, which have some of the properties of water-soluble polymers, gums and alginates from seaweeds are often used. The red seaweeds are particularly valuable as they yield a group of natural mucilages, most of which cannot be replaced by synthetics. One of these is carrageenan, which has both gelling and viscous properties. As well as playing an important role in the food industry, carrageenan, which is often derived commercially from *Chondrus crispus* and *Gigartina stellata* (both known as Irish Moss), is used as a cosmetic thickener.

GUAR GUM

Cyamopsis tetragonolobus

Guar gum is obtained by processing the seeds of the annual forage crop *Cyamopsis tetragonolobus*. The endosperm (plant tissue that provides food for the plant's developing embryo) is milled into a flour which is then reconstituted for use. Pods, produced along a vertical stem like soyabeans, can be harvested with a grain harvester, making collection relatively easy.

The plants are grown extensively in India — where they are thought to have been first domesticated — and Pakistan, and on a large commercial scale in hot, dry areas of the USA, chiefly Texas and Oklahoma.

Approximately 125 000 tonnes of guar gum were produced for world trade in 1987, selling at a cost of about £1500 per tonne in Europe. While much of this was used in paper production, processed foods, textile printing pastes, cosmetics and pharmaceuticals accounted for considerable amounts.

Laminaria spp.

Laminaria or oar weeds, as they are more often known, are brown algae (*Phaeophyta*) belonging to the Laminaceae family. They have large broad fronds with long stems or stipes, attached to the sea bed by multi-branched hold-fasts. Japanese fishermen gather some of the many species of *Laminaria* using long sticks with hooks at the end which they use to twist and tear the algae from the sea bed. Gathering is also done in other areas mechanically, by boat.

The *Laminaria* collected in British waters include *L. hypoborea* and *L. digitata*.

Macrocystis spp.

One of a number of very large brown algae found off the Pacific coast of North America and generally known as kelps, *Macrocystis* species grow together in great quantities and are amongst the largest of all the seaweeds, growing up to 46 metres in length.

The fronds, which can grow as much as 45 cm per day, can be rooted to the sea bed to a depth of 10 to 15 fathoms (18 to 27 metres). Gas filled bladders located near the base of the laminae however, keep the fronds afloat on the surface of the sea.

ALOE
Aloe vera

Aloe vera is just one of over 360 *Aloe* species which flourish in the hot, dry regions of the world. With its long, spiny, pointed leaves which grow in a rosette from the centre of the plant, this succulent is often mistakenly referred to as a cactus. *Aloe vera* is, however, a member of the lily family. When mature, its yellow flowers may tower some 6 metres above the tips of the leaves, on a long central stalk.

Aloe vera is believed to have originated in Africa or Southern Arabia, but it is now distributed throughout the tropical and subtropical regions of the world and may also be found as a house-plant.

Though its use in cosmetics is relatively recent, *Aloe vera* has been used for its medicinal properties for some 3500 years. The Egyptians, Greeks, Romans and Chinese all used the plant for the healing of wounds, skin irritations, rashes, sunburn and headaches, as well as stomach complaints. Alexander the Great is said to have conquered the island of Socotra in order to obtain supplies of aloes for his soldiers' wounds!

When cut, the leaves exude a bitter yellow sap or juice, containing a mixture of phenolic compounds — the source of aloin and aloe extracts. These can be used to help protect the skin from UV rays and are thus incorporated into some modern sun-screen lotions. In addition, the leaf contains a mucilage or gel in its central pulp which consists mainly of a mixture of polysaccharides. It is this gel which has been used for thousands of years for its healing, moisturizing and soothing properties.

The Sonora Indians of Mexico have been using the plant for healing burns and preventing scarring for a very long time, and its modern day usage in this field has extended to the treatment of radiation burns, chronic skin and mouth ulcers, eczema and poison ivy rashes. The gel's special soothing and anti-inflammatory action has highlighted it for use in a range of modern cosmetic creams and moisturizers as well as shampoos. Plants are now grown commercially in large plantations in Florida and Texas.

The larger brown seaweeds, however, all yield alginic acid, from which algin (sodium alginate) is derived. Algin is probably the most useful seaweed derivative to be found in cosmetics, as it can form a base for creams, gels, hair-sprays and colourants — to name but a few! Brown seaweed species commercially harvested for the production of algin include *Laminaria*, *Alaria*, *Fucus*, *Sargassum* and *Macrocystis*. Large quantities of the weed are required for processing with hydrochloric acid and ammonia. Most of the seaweeds we use for algin come from the coastal waters of the USA, Japan and the UK.

Extracts from plants

Price certainly has a lot to do with the plant materials found in our cosmetics today. Since it is not considered economical or easy to use fresh plant material directly — as the Englishman Gervaise Markham did in the seventeenth century, offering a skin lotion which incorporated rosemary, fennel, feverfew, violets and nettles — extracts often obtained by mixing plant matter with a solvent are used instead. Since the range of possible extracts is, in theory, as large as the plant kingdom itself, anything from passion-flowers to oak bark may be used. The only restriction from a commercial point of view is whether the plant matter is available in a suitable form and in sufficient quantity. There certainly seems to be no limit to the number of products — from mouthwashes to sun-tan lotion — in which extracts may appear, though usually in such minute quantities that their efficacy lies in question.

One rather special plant is sold in extract form with claims that elevate it above many others. *Aloe vera* has been soothing, healing and rejuvenating the skin of different peoples around the world for a very long time. Two separate substances have been singled out for special attention: a pale, very pure gel that is contained within the central pulp of the leaves and which is used to heal and moisturize the skin, and a darker juice, which is extracted from the crushed leaves and is useful, amongst other things, for sun-screen preparations. Though some controversy exists over the precise definition and description of the active ingredients, and the efficacy of these substances once processed into creams or lotions for the skin, there seems to be little doubt that in its raw state *Aloe vera* can work wonders!

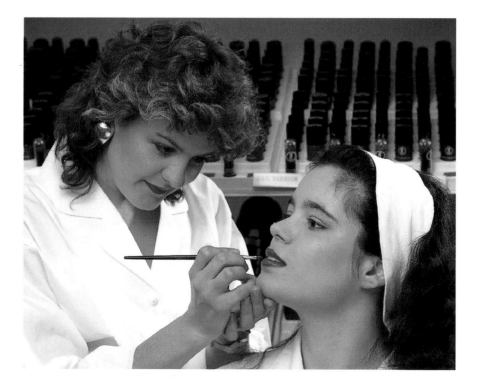

COSMETIC COLOUR — plants that send messages!

If concerned with the texture or softness of our skin, we may well be concerned with its colour. A huge array of cosmetic colours entice us to enhance our looks by emphasizing lips, cheeks, and eyes especially. More coloured waxes, powders, crayons, creams and gels are available to help us do so than ever before.

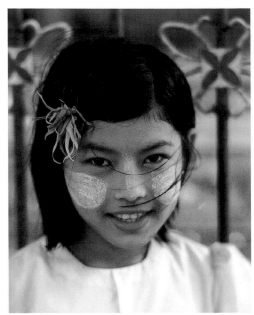

Girl from Rangoon in Burma wearing a face cream made from sandalwood

Tribesman from New Guinea

Japanese lady wearing white face powder

Though we may be amazed at the tribesman from New Guinea with his striking face designs, he wears them to convey important messages about himself and his status. We do the same each time we use eye-liner or lipstick, in even the subtlest shades. Our messages may be much simpler, often concerned merely with attracting people we are likely to meet, but whichever way we view it, colour, or the lack of it is often used to make a point.

Fashion has of course largely dictated the appearance of our faces. Roman ladies were much concerned with whitening the skin and used extraordinary mixtures including pulped narcissi bulbs in the form of face packs. Other face-whitening and highly poisonous ingredients used by European women in the seventeenth and eighteenth centuries included ceruse (white lead), ground alabaster, sulphur and borax.

The rather less dangerous rice flour mixed with plant oils was used as a night-time face pack by aristocratic Chinese ladies, and a face powder based on ground white rice was used during the day. Today most Western face powders are made from complex mixtures including talc, silk cocoons, petroleum derivatives, lanolin, synthetic pigments and various resins. Some, however, may include maize and rice starch in powder form.

Very few of the colouring materials found in modern make-up are plant-derived. Most, like those used by the tribespeople of New Guinea, are based on aluminium or iron oxides which will give a range of colours from pinks and reds to yellows, browns and black. Pressed face powders will often comprise mineral rather than plant oils, but most coloured creams and gels as well as crayons and mascaras are likely to contain the ever-useful carboxymethyl cellulose from pulped wood or cotton linters. As with other products, cellulose acts as a thickener, helping to give a smooth, stable, creamy texture, and in the case of coloured crayons, helps prevent the crayon itself from shattering on application.

Many rouges contain plant cellulose, gum arabic, or algin from seaweed as a suspending agent though the Romans used certain *Fucus* seaweeds to provide the red pigment for their rouge. A similar extract mixed with fish oils is used today by women in Kamchatka, in the extreme east of the Soviet Union.

Lipstick, one of the most popular cosmetics in the West, depends very much on plant materials for its effectiveness. Most lipsticks are based on complex blends of mineral and

RAINFOREST COLOURS

For face and body decoration, many groups of Amazonian Indians traditionally use the juice pressed from the green fruits of *Genipa americana*, a tall stately tree reaching some 18 metres in height. When exposed to the air, the juice from the fruits turns a deep blue-black. Txucarramae women who live in Brazil's Xingu National Park paint some of the most complex designs to be found amongst Brazilian Indians, creating the impression of invisible garments which cling to the body. Charcoal is also used for drawing designs.

Most Amerindians hate drabness and associate it with death. Worn ceremonially, for religious purposes and to show status, they delight in the sensation of colour and design. Red pigments are especially important. In the north-western Amazon a red powder obtained by boiling the leaves of the vine *Bignonia chica* is used by both men and women and is an important item of trade. Chilli pepper juice is sometimes used to help the designs stand out.

The best known red pigment from the Amazon is that obtained from the tree *Bixa orellana*. Known to us as El6O(b) or annatto, a rich paste made from the waxy red coating of the seeds is used throughout Amazonia to decorate the face and body. Whilst we may eat it everyday as the colouring used to liven up our butter, cheese and margarine, the Mehinaku Indians of Brazil apply it to their hair and bodies. Two sorts of urucu, as the paste is known locally, can be distinguished: a deep scarlet used by the men and a paler orange exclusive to the women. Oil from pequi fruits (*Caryocar brasiliense*) which are eaten as a staple food by peoples of the Xingu during the rainy season is used to lubricate the body before urucu is applied.

The Tapirapé, who along with most Amerindians find facial and body hair unattractive, use the fruits of *Streptogyna* grass as tweezers, whilst other wild grasses are used as razors.

The Baka Pygmies, who live in the dense rainforest of South East Cameroon, also use a red pigment to decorate and protect their bodies. It is made from the vivid heartwood of the ngélé tree (*Pterocarpus soyauxii*) which is ground and mixed together with sand and water to make a bright red paste. Rubbed over the bodies of Baka hunters ngélé paste is said to give them courage. It is also used to relieve aches and fevers, give strength to dancers and special protective powers to pregnant and newly-wed women. *Pterocarpus* is used in other parts of Africa too in cosmetics and dyes, for magico-medicinal purposes and in fertility cults.

Red, waxy urucu or achiote paste used in face painting comes from the seeds of Bixa orellana *(below)*

Baka pygmies applying red pigment made from the heartwood of the ngélé tree

vegetable oils and waxes, plus some from animal sources such as lanolin, and beeswax. Plant oils feature in both their whole and fractionated forms: myristic acid from coconut oil for example, is mixed with alcohol to form isopropyl myristate, a colourless liquid softener used to smooth the lips and make them glossy. Fatty alcohols from linseed and soyabean oils are also often present. Castor oil is an important ingredient of lipstick, since as well as having a reasonable taste, it can disperse colour pigments easily and is also fairly resistant to oxidation. Castor and jojoba oils also provide 'slip', helping to spread the lipstick evenly as it is applied.

Plant waxes also play a major part in lipstick manufacture. Two in particular appear repeatedly: carnauba wax, produced by the leaves of the Brazilian wax palm, and candellila wax, which comes from plants of the Euphorbiaceae family. Carnauba wax is a regular if not essential ingredient in lipsticks. A hard wax, it helps to give them firmness and rigidity and is available in various shades and blends. 'Prime yellow', the grade used exclusively by the cosmetic industry, which sells at about £2000 per tonne, is obtained from the younger, fresher palm fronds, whilst the older, tougher leaves will give waxes that progress past corn-coloured to a dark brown or black.

Since the wax melts at only 70°C the various grades are useful for many other purposes. As well as giving Brazil nuts, some confectionery and fruit an attractive shine, carnauba wax has been a traditional component of shoe, floor and car polishes. Produced only in Brazil, a shortage of carnauba wax some ten years ago led to the introduction of candellila wax, which continues to be used together with or instead of carnauba wax. A wax with slightly different properties to carnauba, candellila is obtained by boiling the succulent stems of the shrub *Euphorbia antisyphilitica* in water. The candellila wax used in industry comes exclusively from Mexico.

The brilliant pigments associated with lipsticks come mostly from iron, aluminium or titanium oxides. In Japan, however, the bright red pigment traditionally extracted from the roots of *Lithospermum erythrorhizon* is now being produced by biotechnology, for use in a new range of lipsticks. Forming only 1 or 2 per cent of the dry weight of the roots under normal conditions, shikonin, the red pigment, has been produced since 1983 by growing plant cells in culture. Details of the precise methods remain a commercial secret, but it is estimated that the company concerned has the capacity to produce enough shikonin each year to supply a major proportion of the world market.

Roots of Lithospermum erythrorhizon

Plants as hair dyes

It has never been much fun putting benzine or hydrogen peroxide mixtures on our hair to boost or change its colour. But these and a range of semi-permanent solutions comprising combinations of red, blue and yellow dyes made soluble in water and alcohol have become the standard colouring materials for hair. Almost all the ingredients apart from the detergent bases will be made from mineral oils and ores, but plants are still there to offer us alternatives, in much richer and subtler combinations, if we can only seek them out!

Perhaps the best known natural hair dye in the West is henna. Available commercially as a green-brown powder, it is made from the crushed, dried leaves of the henna plant (*Lawsonia inermis*), and gives a variety of deep red tones when mixed with hot water and applied to the hair in the form of a paste. The colouring material responsible is lawsone or hennotannic acid.

For commercial purposes henna may be altered chemically to provide a base for synthetic hair colourants, but it is also sold in its natural state, sometimes mixed with several other plant materials. These may include indigo, the famous blue-black dye yielded by *Indigofera tinctoria*; ground coffee-beans, or catechu, a reddish-brown dye extracted from the heartwood of the Indian tree *Acacia catechu*.

Two other plant sources have been used since the earliest times to darken the hair; the black

BRAZILIAN WAX PALM

Copernica prunifera

The Brazilian wax palm usually occurs along rivers and on the edges of marshy or saline land in north-eastern and southern Brazil as well as the Chaco region of northern Argentina and Paraguay. It often forms dense forests.

By the time they have unfolded, the young leaves of the palm are coated with a thin glossy film of carnauba wax. Gathered three times a season, the best quality wax is produced by the most tender leaves, some 1300 being required to make just one kilogramme of top class wax.

After gathering, the leaves are graded and dried in the sun for several days. They are then beaten to remove the wax which is processed by importing countries. Carnauba wax is important to a number of industries since it is harder than beeswax and many cheaper, synthetic substitutes.

HENNA

Lawsonia inermis

Henna paste is a popular natural colourant

A shrubby perennial plant which may reach about 3 metres in height, henna is indigenous to Arabia, Iran, India and Egypt. Found growing wild on dry hillsides in these countries and now cultivated commercially in Morocco, China and Australia, the shrub has narrow, grey-green leaves and small, sweet-scented, white or yellow flowers, which are followed by black berries.

Derived from the Arabic name Al Kenna, henna is one of the oldest known hair and body dyes. The prophet Mohammed is said to have used it to redden his beard and it is still commonly used by Eastern women to colour their cheeks, hands, nails and feet. Henna has a special religious and mystical significance for the Berbers of north Africa, who believe it represents both fire and blood and links man with nature. They colour corpses and young babies with the dye and use it in marriage ceremonies, since it is also regarded as a symbol of youth with special seductive powers.

The henna powder sold for use as a hair dye varies greatly in composition and quality, depending on which parts of the plants and which of a number of varieties are used. The finest henna is said to come from Iran. The dried powdered leaves are also known for their cooling and astringent properties and are used as a treatment for leprosy in African folk medicine.

Henna applied to a bride's hands in India

walnut (*Juglans nigra*) and oak galls. Both were recorded by Pliny in about AD 50. One recipe required walnut shells to be boiled with a portion of lead parings and mixed with such exotica as earthworms, to stop the hair from going white, but extracts from the leaves of walnut trees, available commercially today, will do the job just as well!

Oak galls, sometimes known as oak apples, were used by the Greeks and Romans, and by European women in the seventeenth century to blacken hair. Formed by the action of cynipid wasps on the leaf buds of various oak species (including *Quercus robur* and *Q. pedunculata*), they contain relatively high levels of tannin. It was the presence of this tannic acid that promoted their use not just as dyes for hair or cloth but for the manufacture of inks and medicines, and for tanning leather.

For blond hair, rhubarb (*Rheum × cultorum*) and camomile are the traditional herbal lighteners. Extracts of Roman and German camomile (*Chamaemelum nobile* and *Chamomilla recutita* respectively) are those widely advertised in all sorts of modern hair preparations, often in shampoo, but their power to dye the hair is not as strong as that of rhubarb. Containing oxalic acid, rhubarb roots when crushed will yield a yellow dye which can be applied to the hair in paste form to lighten it.

The fresh or dried and powdered roots of English or garden rhubarb (Rheum × cultorum) *boiled in water make an effective yellow hair dye*

Once we have washed or dyed our hair we may well want to hold its style in place. Most hairsprays have traditionally used plant resins in a solvent base to do this job. Though these are mostly now replaced with synthetic polymers, shellac, a dark red transparent resin produced by the action of lac insects on the twigs of the Indian tree *Ficus benghalensis* and other species, may still be used.

SMELLING GOOD!

Universally associated with desire, we have used perfume to make ourselves attractive since the earliest times. It is said that Cleopatra did not win over Antony with her looks alone, but with delicious perfumes which he found himself unable to resist. This same association is used more than effectively today to sell several hundred million dollars worth of perfumes every year.

Though we may think of them primarily in sprays or stoppered bottles on our dressing tables, perfumes are essential components of hundreds of familiar household items. Besides their widespread use in toiletries and cosmetics, perfumes may well have helped us decide which writing paper, disinfectant, disposable handkerchiefs or even shoe-cream to buy!

PLANT SCENTS USED RELIGIOUSLY

Two of the three gifts presented by the Wise Men to the infant Jesus, myrrh and frankincense, were sweet-smelling gum resins produced by plants.

A reddish-yellow in colour, myrrh is produced by a number of small, tropical, thorny trees of the genus *Commiphora* native to Africa and south-west Asia. Some confusion exists over the precise botanical origin of the myrrh so widely used for incense and perfumery in ancient Egypt and the East, but *C. myrrha*, and *C. abyssinica* are important sources. Myrrh is still used today in some cosmetics and perfumes as well as pharmaceuticals.

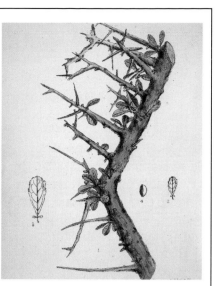

Commiphora myrrha

Frankincense is the gum resin produced by certain *Boswellia* species, including *Boswellia carterii*. It was recommended by Ovid, the Roman love poet, as an excellent cosmetic and was used widely by the Egyptians, Greeks and Romans as an ingredient of incense. When burned, the volatile oils contained within the resin are released and are diffused into the air.

Both frankincense and myrrh were used to anoint the dead in Egypt and formed part of the process of mummification.

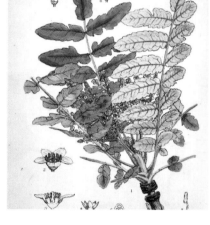

Boswellia carterii

Spikenard was similarly important for religious ceremony amongst the Greeks, Romans and Egyptians. The term appears to have been used to refer to more than one species of plant, however. Whilst spikenard has been identified as a perfume distilled from the desert camel grass (*Cymbopogon schoenanthus*), the 'ointment of spikenard' Mary used to anoint Jesus's feet before the last supper is thought to have been derived from the stems of *Nardostachys jatamansi*, the perennial Himalayan valerian.

Cymbopogon schoenanthus

Hand-coloured copper engraving by Gerrit Valck,
printed by N. de L'Armessin, Paris, 1697

Flowers being used to scent clothes as they dry

The ingredients of many are now made synthetically from various organic compounds, but all perfumes were until relatively recently derived directly from the natural sources (chiefly plants) suggested by their different aromas.

It is impossible to say when mankind first discovered he could extract and use the very varied fragrances of plants to scent his body. One of the earliest recorded means of doing so was by burning scented herbs and inhaling or standing in the smoke. Believed to have special powers because of its smell, and linking medicine with religious ritual, the smoke would drive out evil spirits and associated illness or disease.

Whilst evidence of this ancient use of plants comes down to us in the word perfume, derived from the Latin 'per' meaning through and 'fumus' smoke, the link between perfume and medicine continues today in the form of aromatherapy, which uses combinations of plant oils to correct imbalances in the body.

At least 5000 years ago the Egyptians had developed the art of perfumery to a very high degree. Although pre-dated by the civilizations of China and Minoan Crete, where fragrant plants were probably of great importance, they were already using plant oils such as olive, castor and sesame to anoint their hair and bodies, and these became the bases for infusions of aromatic herbs and gums. Often burnt as incense, and used in the process of mummification, aromatics such as cedarwood, bitter almond, juniper, coriander, henna, spikenard, cinnamon, cassia, and calamus as well as myrrh and frankincense were blended and used as potent perfumes.

An important trade in plant oils and gums developed between Egypt and countries of the Near East, especially Babylonia (present day Iraq) and Arabia. For the inhabitants of all these countries, as for the Greeks and Romans, the use of perfume was not restricted to women. Egyptian men would place a cone of solid perfume on their heads and let it melt slowly, covering the rest of the body.

The Greeks, who attributed the origin of fragrant plants and perfumes to the gods, used different scents for the different parts of the body. Mint, marjoram and thyme as well as fragrant flowers such as rose and hyacinth were favourite ingredients of some cosmetic oils and scents.

The Romans, even more lavish in their use of perfumes than the Greeks, scented not only their bodies and hair, but their clothes, bedding, military flags and the walls of their houses! Saffron, cardamom, costus, melissa and narcissus flowers were just some of the ingredients used.

FRAGRANT GARDENS

One of the oldest recorded collections of aromatic plants — thirty-one frankincense trees, carried in wicker baskets from Somalia to Egypt — was planted for Queen Hatsepshut in her temple on the banks of the Nile around 1500 BC.

A tradition maintained by the Pharoahs, the last Egyptian ruler Cleopatra (69 to 30 BC), owned a 'balsam garden' containing aromatic plants and trees whose oils and resins scented her hands and body.

At about the same time, it has been suggested that the biblical Garden of Gethsemane could have been a jasmine garden — if Gethsemane were taken to be a mis-translation of 'Jessamine', indicating a place where jasmine, used to make a fragrant oil was grown. Still a popular ingredient in perfumes today, jasmine has long been a favourite of Eastern countries and in China the flowers of *Jasminum sambac* are used for scenting tea.

By the seventeenth century, sweet-smelling herb and nosegay gardens, begun in the monasteries of southern Europe and influenced by Roman villa gardens, had come to occupy an important place in many European countries. They supplied medicinal and culinary plants to the great town and country houses, as well as the ingredients for perfumes and scented waters. Elaborate knot gardens became especially fashionable. They were planted with aromatic herbs spaced well apart, in lines or groups dictated by the geometric shapes. Some of these have survived into the present century.

A fragrant garden at Chenies Manor, Hertfordshire, England

Whilst the trade in gums and plant oils continued to flourish in the Holy Roman Empire, it was not until the late tenth century that the Arabian doctor Avicenna invented or perhaps rediscovered the art of distilling essential or volatile oils to make essences and aromatic waters. The fame of his rose water in particular spread and reached Europe along with other exotic perfumes from the East at the time of the Crusades.

By the end of the twelfth century Europe had its own perfumers and by the thirteenth, when lavender was already being grown at Mitcham in Surrey for this purpose, it was developing distinctive fashions of its own. The process of distilling oil from plants and flowers in alcohol gave rise in the fourteenth century to Hungary and later Carmelite waters — the latter including angelica, lemon peel, lemon balm and coriander in a base of orange flower water and alcohol. By the sixteenth century, most large houses in Europe possessed a stillroom, where herbs and flowers were processed into simple perfumes.

Eau de Cologne from Germany, one of the most famous of perfumed waters and still very popular today, was invented at the beginning of the eighteenth century. Its success was due to a combination of rosemary, neroli and bergamot oils distilled in a grape spirit.

None of the plants or flowers so valued in the past or those prized by the perfume trade today would have been used at all if it were not for their essential oils. Also known as volatile oils because of their ability to diffuse into the air, they are the essence of a plant's fragrance. Not all plant oils of course, smell as good to us as others, though very different plants will sometimes release quite similar odours. In nature, the perfumes given off by flowers are an important means of attracting pollinating insects. Some, such as that produced by the maroon and mustard coloured flowers of the succulent *Caralluma speciosa* which mimics the smell of rotting meat, are attractive to flies. Those with paler flowers — since their perfume replaces and is derived from pigments similar to chlorophyll — have delicious scents that we have come to cherish. Lily of the valley, honeysuckle and ylang-ylang are just three examples of these.

Essential oils may be present in different combinations and concentrations in different parts of a plant. Whilst orange flowers contain their own distinctive oil known as as neroli, orange peel produces a citrus oil. Lemon grass oil on the other hand, comes from the leaves and stems of a lemon-scented grass, cinnamon oil comes from a bark, and sandalwood, pine and cedar wood oils are extracted from the wood of these trees.

Various methods are used to extract essential oils depending mainly on their location within the plant. Citrus oils can be literally pressed from the skin of the fruit, but delicate flowers such as jasmine and tuberose must receive much gentler treatment. In a process known as enfleurage, these blossoms are spread thickly between shallow trays greased on both sides with purified fat, or in cloths soaked in olive-oil, and left to impart their fragrance. The flowers are removed and replaced with fresh ones every few days until the fat or oil has become saturated. The essential oil is then extracted with an alcoholic solvent.

One of the most common methods of extracting essential oil, however, is steam distillation. This involves the heating of plant material by forcing steam through it under pressure. Tiny particles of essential oil are vapourized and carried away in the steam which collects in a condenser. Once cool, and having separated, the oil can simply be skimmed from the surface of the water which has formed. Though the idea is simple, the process is a highly skilled and delicate one. Most essential oils have an extremely complicated chemical make-up — several hundred different types of molecules might be present in each one. To heat and subsequently alter this chemical composition may change its odour, so only the tougher flowers such as roses or lavender, or the leaves, bark, seeds or roots of certain plants may be suitable.

A gentler heat is employed in the process known as maceration. Here successive batches of fresh flowers are left to soak in warm fat for several days until the fat becomes impregnated with their essential oils.

Solvent extraction avoids the use of heat altogether. Plant material is kept constantly

ROSE

Rosa spp.

Amongst the oldest and most famous of perfume ingredients, oil or attar of roses continues to play a major role in perfumery and aromatherapy. Its particular properties make it one of the most antiseptic of the natural oils and one of the least toxic. Rose oil has long been valued for its soothing action on the nerves, which may induce sleep, as well as for its powers as an antidepressant. It is also said to tone the vascular and digestive systems.

According to legend, rose oil was discovered in Persia (now Iran) when a princess at her wedding feast noticed the oil released from rose petals which had been thrown into a pool and warmed by the sun. Persia was once very famous for its rose water, but the production of the rose oil that gave rise to it, is no longer commercially significant. India was also a major supplier of rose oil in the past, but Bulgaria and Turkey have taken its place, producing the best quality and therefore most expensive oil, from the Damask rose, *Rosa damascena*. Other rose species used for this purpose include *R. gallica*, cultivated widely in Morocco, and *R. centifolia* grown in the Grasse region of southern France.

Rose oil is not red in colour but an orange-green. Some 60 000 roses are needed to make just one ounce of oil!

Rosa gallica

Perhaps the most popular plant in Europe, there are some 200 to 250 different species of rose. Those mostly found in our gardens are the hybrid teas derived from the tea rose (*Rosa odorata*) and the floribunda species. The wild dog rose (*R. canina*) which graces our hedgerows, is also a distant ancestor of the hybrid tea.

YLANG-YLANG

Cananga odorata; Artabotrys odoratissimus

One of the key ingredients used by French perfumers, the oil of ylang-ylang has a voluptuous scent and is distilled from the beautiful yellow flowers of a tall tropical tree, *Artabotrys odoratissimus*. It grows in Madagascar, Java, Sumatra, Réunion and the Comoro Islands, though the best oil is said to come from Manila, in the Philippines.

The production of oil is a very delicate business. The flowers, which are hand-picked in the early morning, must not be bruised or they will turn black and start to ferment. The oil also requires expert handling once extracted and is a very precious commodity.

The flowers of *Cananga odorata*, closely related to *Artabotrys odoratissimus* are also used to produce an oil known as ylang-ylang, but its sweet, exotic scent is said to be slightly inferior.

immersed in alcohol or volatile solvents such as petroleum ether and is then distilled to leave the oil behind. The primary substances obtained from this method and those of enfleurage and maceration are known as 'concretes' and are considered to be the purest of the natural fragrances. Since they will contain waxes and other plant materials, the essential oils can be further separated out with solvents to leave 'absolutes'.

The majority of perfumes consist of mixtures of these absolutes; rose, jasmine, and orange are some of those that frequently occur. Perfume then, is not a single smell but a combination of odours that will have been chosen for their compatibility. Very few perfumes have less than twenty to thirty ingredients and many have over one hundred. As well as the natural plant odourants — of which there are about three hundred — animals provide some very important substances. Civet from the civet cat, castoreum from the beaver and musk from the musk deer are often used as 'fixatives', helping to slow down the evaporation of the volatile oils and so 'fix' their odours. Some plant resins also perform this function.

Eau de Cologne from Germany is one of the oldest of the scented waters and is still popular today

Blooms from white dove orchids grown in the Malay Peninsula being processed for oil extraction

Cypress oils being distilled

Though the more expensive perfumes rely to a great extent on these natural plant and animal sources, which are often very difficult to imitate exactly, some 60 per cent of all the fragrant materials used today are made synthetically. Many of these synthetics, however, are still derived from plants.

As well as chemically isolating some of the individual components of essential oils to make separate substances — geraniol for example, from citronella and palma rosa oils — copies of traditional ingredients and many new ones are made from quite separate plant materials. Turpentine, extracted commercially from several species of pine tree is a major source of

SANDALWOOD

Santalum album

Once common in the dry regions of peninsular India, especially the states of Mysore and Tamil Nadu, the sandalwood tree has played a prominent part in Indian culture for some twenty-three centuries.

Known in commerce as East Indian sandalwood to distinguish it from other *Santalum* and unrelated species which are used as substitutes, the tree, which may grow up to 18 metres in height, is evergreen and semi-parasitic, using nutrients from the roots of other trees to help it grow.

One of the few woods which is weighed correct to fractions of a kilo and formerly classified into eighteen different grades, sandalwood timber is extremely valuable. Great quantities have been used for buildings and for carving ornaments and both the scented red wood and the oil are traditionally used in incense and medicines.

The most famous use of the oil however, is in perfumes. Produced mostly by steam distillation of the chipped wood (from the roots and heart of the tree), it has a heavy, sweet scent and a very lasting odour. Since it has no overwhelming top notes however, and with the advantage of its pale colour, it is very valuable as a fixative and can be blended with other oils, such as rose, to make attars.

Ninety per cent of this warm, spicy oil is now used in soap, perfume and cosmetics. In India, Hindu women apply a paste made from the ground wood to their foreheads, thus combining religious and cosmetic uses.

Overharvesting of sandalwood has now led to a widespread shortage and higher prices. Though substitutes from other parts of the world are used, *Amyris balsamifera* for example, known as West Indian sandalwood and *Osyris tenuifolia* from tropical and South Africa, nothing is said to compare with the fragrant qualities of East Indian sandalwood.

Santalum album L.

aromachemicals. Flavours and fragrances as diverse as spearmint, nutmeg, lavender and lime can be made from the pinenes in turpentine, as well as valuable compounds such as linalool, an important component of certain natural and synthetic oils.

Putting all these sources together, the modern perfumer may have up to one thousand ingredients at his disposal to create a masterpiece! Having done this, his perfume will be distinguished by its 'notes'. 'Top' describes those odours most immediately detectable, 'middle', those that give the perfume its intrinsic character, and 'base' its fixatives. Notes may range from the 'floral' — rose, hyacinth, or lilac for example — to 'green' — referring to oak moss, lichens, ferns or ivy — or even 'maritime', denoting seaweed!

The French, whose perfumers were granted a special charter as early as 1190, are still considered leaders in the perfume trade, despite much competition from America. Of the two to three million dollars now needed to launch a major perfume, only a very small part will relate to the raw materials used, some of which may still be grown or gathered locally on a very small scale.

At the end of the day however, whether using turpentine to manufacture the smell of fresh limes for commercial use in aftershave, or rose petals at home to make a simple scented water, plants leave us smelling and often therefore feeling a great deal better than we would without them!

Keeping us covered

FROM THE FIG LEAF

If we believe the Bible, man's first item of clothing was literally plucked from a plant. The fig leaf must have proved a somewhat precarious covering for Adam's nakedness, however, and it is hard to imagine it being used in this way for very long.

The steady observation of plant materials and subsequent experimentation appears to have led our ancestors to a much more skilful and ingenious use of plants very early on in our history. Many societies soon discovered that it was the fibres, either external or internal, separated out or processed in various ways that were by far the most useful parts of plants for making clothes. Cotton, for example, which came to be cultivated in the Old and New Worlds independently — and which is our most important natural fibre — was being grown for use in coastal Peru as early as 10 000 BC.

Other societies developed simpler but none the less effective ways of covering their bodies. The Yagua Indians of Colombia still use the dried and split fronds of jungle palms to make fringed items of dress, and almost all rainforest peoples have an extensive knowledge of the uses that fibres from a number of palms and other plants can be put to for a variety of body ornaments.

'GRASS' SKIRTS AND BODY ORNAMENTATION

The Yagua Indians who live in the north-western Amazon region of South America have always used the leaves of the Chambira palm (*Astrocaryum chambira*) to make their multi-fringed clothes. The petioles or leaf stalks of the freshly gathered leaves are soaked in water and then split to separate the fibres into long strands which can be knotted or threaded into place.

Many other palm species native to tropical America have provided fibres for body apparel, including arm and leg bands, necklaces, amulets and bracelets. Chambira palm fibres are generally regarded as the finest for weaving, but those obtained from the buriti palm (*Mauritia flexuosa*) and various *Bactris* species are also important and are commonly made into hammocks, nets and multi-purpose cords.

Some groups use penis sheaths made from woven palm leaflets whilst the nuts produced by *Astrocaryum* palms are carved into ear-rings and beads for necklaces. Many tropical seeds and sometimes carved scented woods are also used as beads, often accompanying animal teeth and feathers in very beautiful and eye-catching combinations.

Leaf skirts in Hawaii

---

*Fibres from the buriti palm (*Mauritia flexuosa*) are used for weaving in South America*

In Papua New Guinea many of the 700 different tribal groups use the stems and seeds of pit-pit grass (*Coix lacryma-jobi*) for decoration. Whilst Huli men wear the stalks through their noses, Henganofi women cover themselves in 'necklaces' made of the seeds to show that they are widows. Sometimes weighing up to 23 kg in total, the strands are removed one by one over a span of several months until the official period of mourning is over.

Cordyline leaves known locally in pidgin as 'arse-grass' are worn decoratively by some tribal men to cover their buttocks. These leaves are generally suspended from multi-stranded cords of twisted fibres or strips of painted bark cloth. Long, curly bean gourds are sometimes worn as penis sheaths.

Woman in Papua New Guinea wearing beads made from the seeds of the pit-pit grass (Job's tears), as a sign of mourning

PLANT FIBRES FOR CLOTHES

One of the most interesting and extensively used plant materials for clothing, but which never became an item of world trade, is bark cloth. Captain Cook's first voyage to Tahiti and the Pacific Islands brought back news and actual examples of a cloth 'almost as thin as muslin' and often dyed beautiful shades of red, made from the beaten inner bark of certain trees. Sir Joseph Banks, naturalist on this voyage, recorded the materials and processes involved in the making of this 'kapa' or 'tapa' cloth (from 'ka' the and 'pa' beaten — meaning 'the beaten thing'), in some detail.

In Tahiti he observed the making of various sorts and thicknesses of cloth from a number of different trees, which were grown and carefully tended for the purpose. The most important included the paper mulberry (*Broussonetia papyrifera*), the breadfruit tree (*Artocarpus altilis*) and certain species of fig. The bark of these trees was stripped from trunks having grown only to 2 or 3 inches (5 to 8cm) in diameter, and placed in running water for several days. The outer layer was then scraped with pieces of shell until only the fine inner fibres remained and these would then be spread out in layers for beating.

Joseph Banks related in his journal:

> They laid them in two or three layers, and seemed very careful to make them everywhere of equal thickness, so that if any part of a piece of bark had been scraped too thin, another piece was laid over it, in order to render it of the same thickness as the rest. When laid out in this manner, a piece of cloth is eleven or twelve yards long, and not more than a foot broad, for as the longitudinal fibres are all laid lengthwise, they do not expect it to stretch in that direction, though they well know how considerably it will in the other . . . It is then taken away by the women servants, who beat it in the following manner: they lay it upon a long piece of wood, one side of which is very even and flat, this side being put under the cloth: as many women then as they can muster, or as can work on the board together begin to beat it.

The women used rectangular wooden batons with grooves of differing widths cut into each of the four faces. The cloth would be beaten until it reached the desired thinness, often being doubled over and splashed with a glutinous liquid containing either the mucilage from hibiscus plants, taro or arrow root starch to help the fibres adhere, then bleached and often dyed.

Tongan woman painting tapa bark cloth

THE PAPER MULBERRY

Broussonetia papyrifera

Cultivated since the earliest times in Japan, China and Polynesia, the paper mulberry is perhaps the best known source of fibre used for making bark cloth. Unlike cotton and kapok which are 'surface' fibres, the paper mulberry is used for its internal stem or 'bast' fibres. These fibres which can be obtained from several other dicotyledonous plants are made up of long strands of cells, held together by gums and pectins. They are very strong and unlike those from leaves, known as 'hard' fibres, can tolerate bleaching.

In China around AD 100, fibres from the paper mulberry's bark were separated and mixed with those of flax and hemp to make a form of paper. The Mayans independently developed a way of making paper-like sheets by pounding the bark of various trees. They used this not as the Polynesians did for clothes, but for recording by means of hieroglyphs, their customs, traditions and cosmology.

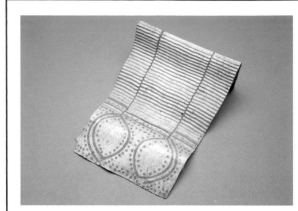

Bark cloth from the Amazon crimped with the teeth

Amazonian funeral dress made chiefly of fig bark, worn by Cubeo Indians (of the Vaupés river)

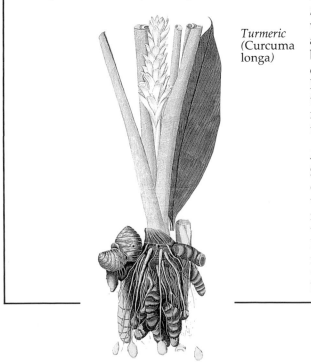

Turmeric (Curcuma longa)

CLOTHES AND COLOUR FROM FIGS

The fig trees that Captain Cook saw growing on Tahiti, along with the paper mulberry and the breadfruit, all belong to the Moraceae family.

Within this family, the genus *Ficus* is particularly interesting. It contains over 800 different species of fig and many of these have supplied peoples living in the tropics with fibre for cloth as well as raw materials for dyes.

Once commonly worn by Polynesian islanders, bark cloth made from fig species is still used by many ethnic groups of the Philippines, Malaysia, Indonesia and Papua New Guinea. Some Amerindians make items of clothing and slings for carrying babies from fig bark and its use has been traditional in Madagascar and parts of Central and Eastern Africa, where the baobab (*Adansonia digitata*) is also used. In the forests of Cameroon, the Baka pygmies traditionally used *Ficus* and other tree species for making cloth.

Around the world all manner of bark cloth garments have been made — from jackets worn by the Dyak people of Borneo to the trailing robes of Samoan chiefs. The ultimate in cool clothing for hot climates, once dirty or worn-out, bark cloth can be washed, repulped and beaten again to make clothes that are as good as new.

The dye material produced by some species of fig comes not from the bark but the sap. Joseph Banks found the red dye used by Tahitian women for staining tapa cloth 'most beautiful, I might venture to say a more delicate colour than any we have in Europe, approaching, however, most nearly to scarlet'.

The dye was made by mixing the milky sap exuded drop by drop from the fruit stems of tiny figs (*Ficus tinctoria*) with the leaves of various tropical trees and plants such as *Cordia sebestena*, *Solanum latifolium* and *Convolvulus brasiliensis*, which were first placed in 'cocoanut water'. The action of the fig's sap on the leaves produced a beautiful, if transitory dye. Banks, at pains to describe it exactly added: 'The painter whom I have with me tells me that the nearest imitation of the colour that he could make would be by mixing together vermillion and carmine, but even thus he could not equal the delicacy'.

More plant dyes for bark cloth

Still used in Papua New Guinea for dying bark cloth, turmeric (*Curcuma longa*) is traditionally valued in many parts of Asia as a yellow dye. The roots of the plant are scraped and mixed with water and the cloth is left to soak in this mixture. Brown dyes can be obtained from mangrove bark, rich in tannins, while cloth is still dyed black in parts of South East Asia, by burying it in mud beneath plots of taro (*Colocasia esculenta*).

Banks and the scientists and explorers who succeeded him found evidence of tapa-making — using a number of different tree barks — all over Polynesia, from Hawaii to Papua New Guinea. Sometimes made in pieces that were hundreds of feet long, red and yellow plant dyes were much favoured for its decoration and the cloth was often glazed or varnished with a special vegetable gum, then stamped with abstract or leaf designs. Both men and women wore tapa cloth, styles ranging from elaborate wraps and poncho-like garments to simple loin cloths or 'skirts', according to region and rank. Tapa was also used to make mosquito curtains and fine screens as well as bedding.

Banks, impressed with the cloth's softness which he said resembled that of 'the finest cottons', often slept in it, finding it 'far cooler than any English cloth'. With the influx and availability of manufactured Western clothes and the ever-present pressure to conform, tapa-making has now been abandoned in many areas, but some islands including those that make up the Kingdom of Tonga continue to practise this very ancient art for ceremonial occasions and the tourist trade.

Another very fine cloth made not from stem but leaf fibres was worn traditionally by the Maoris of New Zealand. Greatly impressed with the quality and softness of this clothing, Captain Cook noted that it was made from 'a grass plant like flags, the nature of flax or hemp, but superior in quality to either'. The plant was in fact New Zealand flax (*Phormium tenax*), not a grass but a member of the lily family. Still used today by some groups, by the time of Cook's arrival the Maoris had developed a large number of uses for the very strong fibres, which were separated from the narrow leaves — sometimes growing to a height of 2 to 2.5 metres. The many different varieties of the plant offered fibres suitable for different purposes. As well as several thicknesses of twine and fishing line, both coarse and fine matting was made and many different types of fabric.

To separate the fibres from the leaves — selected with great care for the finest cloth — the Maoris scraped them with mussel shells and then soaked them in water, drying them subsequently in the sun. This technique is equivalent to the current commercial processes of 'decorticating' — the scraping of non-fibrous material from plant fibres, and 'retting' — the removal of soft plant tissue containing gums and pectins from the harder fibres by letting them decompose.

Maori war canoe. c. 1770

Once separated out, the Maoris knotted the pure white fibres together to produce very strong cloth, some of it as fine as silk. Today New Zealand uses its native flax plant for the commercial manufacture not of cloth, but very hard-wearing floor coverings, underfelt for carpets, upholstery materials, furniture padding, and different sorts of rope and twine.

Clothes fit for a king!

Neither the soft bark cloth of the Polynesians nor the silky fabrics made by the Maoris were adopted for use in Europe after their 'discovery', but the delicate inner fibres of the lace bark tree (*Lagetta lagetto*) met with greater success.

Held in several layers beneath the outer bark of the tree, this entirely natural network of fibres which can be stretched apart to resemble the finest hand-made lace had traditionally been used by the peoples of Jamaica, Cuba and Hispaniola where it is native. With the arrival of the Spanish, and at their instigation, it was made into bonnets, capes and entire lace suits for Europeans!

The most famous use of lace bark, however, was a cravat, frill and pair of ruffles sent from Jamaica in the seventeenth century and worn in Britain by King Charles II!

Raffia costume being worn during an 'Epke' religious masquerade in West Africa

Raffia weaving

Raffia for keeping cool

In many parts of tropical and sub-Saharan Africa the fibres of the raffia palm have been used for centuries to make traditional clothing. Though this use has decreased with the introduction of European cotton cloth, raffia is still widely worn in West Africa and on the island of Madagascar, especially for ceremonial occasions. The most splendid and elaborate raffia costumes were worn by people of status such as local elders and ritual specialists at masquerades and dances, and denoted the power and prestige of the wearer by their colourful design.

Believed mistakenly by colonial visitors to Africa to be a grass, raffia is in fact a palm leaf fibre. There are twenty species of raffia palm world-wide and all of them are found in Africa with one species in South America and one in Madagascar. Raffia palms have the biggest leaves in the entire plant kingdom. The stemless *Raphia regalis* which grows in Cameroon can produce leaves that are up to 25 metres in length! One of the most widely used palms in West Africa is *R. vinifera*, whilst in Madagascar, *R. farinifera* is of supreme importance to local people for a range of domestic purposes.

Slightly different methods are employed in different regions to produce the supple, straw-coloured strands still coveted by European gardeners, but the general principles are the same. Young and tender palm leaves are first detached from the petiole of the palm frond and whilst they are still fresh a small incision is made on their underside with a knife, close to the top. The raffia fibre, a colourless membrane produced by the upper epidermis of the leaflet, is then quickly pulled away by hand or by running the leaflet across the blade of a knife. The membranes are next tied together at one end — about ten at a time — and left to dry in bundles in the sun, during which time they will acquire their familiar creamy yellow colour. For good quality fibre that does not roll up at the edges to look like string, the drying process must be carried out very carefully. The fibres, which can be split into very fine strands if required, are now ready for knotting, knitting or weaving together to make costumes.

In the Zaïre Basin area raffia was the only plant fibre traditionally woven into cloth, but its production has now decreased. The fibres from the raffia palm remain an important source of rope, fishing tackle and all sorts of cord, however, and the midribs of the leaves are often used as roofing poles. An alcoholic wine is also made from the sap which is tapped from the trunk of the palm.

It seems likely that the early invention of techniques for the spinning and weaving of certain natural fibres into cloth had much to do not only with the availability of these materials, but with the climatic conditions that prevailed and lifestyles of the peoples involved. Whilst it is often pointed out that groups such as the Polynesians 'did not know the loom', as if this were the sign of a flaw in their 'development', it is perhaps forgotten that the materials they used and the ways in which these were worn were determined by both practical and important socio-religious considerations. The same may be said of many Amerindians who despite their knowledge of the spinning and weaving techniques which Western-style clothing now largely depends on, devised the means of making appropriate outfits of their own.

FIBRES TO WOVEN FABRICS

Few of us think much about the raw materials that go into our clothing today, or why we are using them, but almost all of us rely on plants to a greater or lesser degree. They supply us with many types and textures of cloth, including a number of man-made fabrics which actually start out as trees!

Pure cotton is a very popular natural fibre

Cotton plant
(Gossypium barbadense)

Cotton

The chances are that each of us is wearing at least one garment made from cotton, the most important natural fibre in the world today. Surprisingly few plant fibres have been used

*Cotton being taken into ginning room in
Acurangabad, India*

COTTON

Gossypium hirsutum

Cotton comes from the same family as hibiscus
and the hollyhock (the Malvaceae). With over
thirty species in existence native to parts of Asia,
Africa, South and Central America and Australia,
the flowers of the cotton plant may range from
creamy white to beautiful reds and purples.

The flowers of *Gossypium hirsutum*, which
provides most of the world's cotton, are a pale
cream colour. In tropical countries cotton can be a
perennial, developing thick woody stems and
roots, and some wild species grow into small
trees. In the southern United States, which has a
climate too cold for the plants to over-winter,
cotton is sown from seed each spring. After the
pollination of the flowers a large fruit or boll
develops, containing about ten seeds. It is the fine
hairs attached to these seeds, revealed when the
ripe boll bursts open, that are the cotton fibres of
commerce. Once picked by hand, chemicals are
now used to defoliate the cotton plants, leaving
the bolls, resembling ragged snow-balls, to be
gathered by machine. The fibres are stripped from
the seeds in a cotton gin before being packed into
bales.

One kilo of cotton contains roughly 200 million
individual seed hairs — each one a single cell,
3000 times longer than its width!

Cotton weaving

Some of the finest and most beautiful of all woven
cotton was produced in Peru.

A succession of coastal and highland cultures
brought spinning and weaving to such a fine art
that weavings continued to be the most prized of
possessions and most sought-after trading
commodities up to the time of the Spanish
conquest.

Almost all the weaving techniques known to us
today were known to the ancient Peruvians. They
produced an extraordinary variety of fabrics from
simple plain weaves to intricate gauzes, fringed
brocades, open-work, tapestries and cloth that
could be tie-dyed, printed on or painted.

Mummies preserved in the dry sand of the
coastal desert were sometimes wrapped in
thousands of square feet of cotton cloth. This
fragment of mummy cloth was woven by people
from Chancay, just north of Lima, whose culture
flourished between 100 BC and AD 1200. Natural
white cotton cloth has been painted with a red-
brown dye to reveal characteristically geometrical
seabird and feline designs.

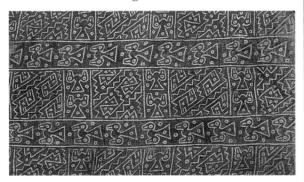

*Cotton mummy cloth of the Chancay dynasty, c. 1000
years old*

Cotton production

The Peoples' Republic of China is the world's
biggest producer of cotton. An estimated 4202
million tonnes was grown there in 1986-87. After
China comes North America and the USSR,
which both grew over half this amount. India,
Pakistan, Egypt, Brazil and Turkey are also major
producers. Japan was the individual country to
import most cotton last year, whilst Europe and
America each consumed around one and a half
million tonnes.

The rise and fall of cotton

It is interesting to reflect that in the sixteenth century many Europeans still thought that cotton came from a kind of tiny 'vegetable lamb' that grew on trees! The first tales of cotton cloth came not from the New World — where Columbus and later explorers were indeed amazed to find exquisite fabrics being worn and used for hammocks — but from travellers to the East. Accustomed to much coarser garments made from linen or wool, stories of the fine fabrics of India, China and Japan fuelled feverish imaginations.

The word cotton itself comes from the Arabic equivalent 'qutun'. The Arabs had long been growing plants native to the Old World and trading in cotton fabrics, and the term muslin comes from the Arabic name Mosul, a town in Mesopotamia, where this smooth and delicately woven cloth was first made.

It was Indian cottons, however, that were the first to reach Europe on a large scale — the Venetians set up a monopoly and organized camel trains to bring the cloth to Italy. To avoid the necessity of importing cotton from India themselves, the British set up cotton plantations in the Caribbean and the southern USA with seeds from Indian plants. In 1732 they also introduced the upland cotton (from seeds grown in London's Chelsea Physic Garden) which was to form the basis of all the cotton grown in the USA.

With the later eighteenth century inventions of new machinery for the carding, spinning and winding of fibres, and a cheap and plentiful supply of cotton from America (where the invention of the cotton gin had dramatically increased output), the British industry took off.

By the early 1800s Liverpool had become the centre of the world trade in cotton and by 1850, 60 per cent of all British exports were cotton cloth. However, the inequalities which were to develop between the northern and southern states of America, with southern farmers constantly indebted to the northern bankers who marketed their crops, and disputes over trade policy, led eventually to civil war and the subsequent decline of the British cotton industry.

commercially to make clothing; cotton is certainly one of the oldest and best known. It is also currently the world's most important non-food plant commodity.

The fibres come, not as we have already seen from leaves, stems or bark but from the seeds of the cotton plant. Attached to the outer surface of the seeds each cotton fibre, which has the appearance of a fine hair, is one single seed coat cell. Beneath these long fibrous hairs lies a second layer of shorter, fuzzy fibres, known as linters. Removed from the seed these are used for making water-soluble polymers and paper.

The biological function of the cotton seed's silky hairs — also known as 'surface fibres' — is to help the seeds disperse in the wind. The strength and flexibility of these fibres however, plus a gradual spiral in their length which causes the strands to interlock as they are twisted, has made them very suitable for spinning into thread.

Evidence suggests that one species of cotton, *Gossypium barbadense*, was used as early as 8000 BC by the peoples of coastal Peru, and domesticated there by 2500 BC. It appears, however, that cotton cultivation and spinning and weaving techniques developed quite independently in different parts of the world. Archaeological discoveries indicate that its first use in the Old World was in south central Asia, and pieces of cloth have been found in Pakistan that are 5000 years old.

When it comes to the history and taxonomy of cotton species the picture becomes very complex. Of the thirty recognized species only four are now cultivated on any scale commercially, but each of these has numerous subspecies, varieties and cultivars.

There is still no firm agreement amongst botanists about the ancestors of these four, but it is generally accepted that *Gossypium arboreum* and *G. herbaceum* were cultivated in the Old World whilst *G. hirsutum* and *G. barbadense* developed in the New. To complicate matters, however, both these New World cottons appear to have one New and one Old World parent: *G. raimondii* endemic to northern Peru and *G. herbaceum* from Africa.

Since they have longer fibres or 'staples' than their Old World cousins, it is the New World species that supply us with almost all the cotton grown commercially today. *Gossypium hirsutum* which accounts for about 95 per cent of this, is thought to have originated in north Brazil, spreading both east across the Amazon and over the Andes, and northwards to central America. It may well have been domesticated in Mexico and was certainly being grown there some five and a half thousand years ago.

The Aztecs were accomplished weavers and the Mayans so revered the art that it was given its own patron: the moon goddess Ixchel. Only Mayan nobles and their servants were allowed to wear fine cotton clothes though, the ordinary people wore garments made of maguey fibres, extracted from the tough leaves of *Agave pacifica*.

Most upland cotton, as *Gossypium hirsutum* is also known, is grown today in the USA and Russia. Its cultivation has been favoured partly because of its greater resistance to boll weevil attack than some other species. The boll weevil came originally from Mexico, where it was a minor pest on wild cotton, but the commercial production of other species helped it to spread by the beginning of this century to cover the entire southern USA — reducing yields in some areas by as much as 50 per cent. The subject of continued research for chemical compounds to eliminate it, the weevil still causes the loss of some 5 per cent of the cotton crop each year.

The more exotic-sounding sea-island Egyptian or Pima cotton is in fact *Gossypium barbadense* native to northern Peru. By the time Columbus arrived in the West Indies, it had already been introduced there by native people and was later taken by Europeans to plantations in the southern United States. Cultivars of this species — selected for their very long fibres, sometimes 6 cm in length — were also introduced to Egypt and further selections produced Pima cotton, named after the county in Arizona where they were once grown. *G. barbadense* produces fibres for most of the high quality 'luxury' cotton cloth sold on world markets today.

Sizing up the potato!

Cotton is the most popular natural fibre in the world. The characteristic garments of many different peoples are traditionally made from it, and in our industrialized societies, we have come to associate the 'pure cotton' label with comfort and style.

Before much of our contemporary cotton clothing can be made, half of the cotton threads involved will have come into contact with some important substances produced by other plants. Once they have been spun, the fibres which will make up the warp threads of almost all cotton and some synthetic fabrics must be 'sized' (stiffened) to make the weaving process easier. Only the fine long-staple sea-island cottons do not need to be treated with the glue-like size solution which will prevent the warp fibres (placed lengthwise in the loom) from being pulled apart during the rapid weaving processes, without losing elasticity. The size solutions used are often made from natural starch.

Though it is present in almost all plant parts, starch tends to be stored in the largest quantities in various seeds and tubers. In Europe the potato is a major source, Holland being a principal producer, whilst America relies heavily on maize and wheat grains. Other plentiful sources of starch which are used commercially include cassava tubers (*Manihot esculenta*), from which tapoica pudding is made; arrowroot (*Maranta arundinacea*); taro tubers (*Colocasia esculenta* var. *antiquorum*); sorghum grains (*Sorghum bicolor*) and the pithy stems of the sago palm (*Metroxylon sagu*).

The starch itself (a natural polymer made up of glucose molecules) is present within the plant cells in the form of insoluble granules. When the cells are crushed, for example when a potato is cut or peeled, then placed in water, the starch is simply washed out and the water takes on a characteristically milky appearance. If the water is then evaporated a solid starch residue is left behind, which can easily be dried and powdered. Cornstarch, arrowroot and potato starch are all sold in this form for domestic use.

As all gravy experts will know, the starch powder must be mixed with cold water or a water-based solution and then heated before it can be used. The presence of heat causes the starch granules to expand and absorb the water, which results in a thick solution or gel. For use as a size, starch may be treated with heat in this way or with alkalis, oxidants or enzymes to produce a range of gelatinous liquids.

The type of size used depends to some extent on its price and the fabric to be treated. Cotton, for example, can be adequately sized with potato starch which is relatively cheap. Polyester on the other hand may be sized with a mixture of potato starch and a 'synthetic' polyvinyl alcohol.

Once the material has been woven, and before it can be worn, the size must be removed. Since potato starch is not strictly soluble in water, forming an opalescent dispersion rather than a true solution, washing the sized material in water does not dissolve it away. To remove it completely the material must be brought into contact with carbohydrate-eating enzymes which break down the starch into sugars or smaller molecules. This process is very expensive though, and as a general rule the cheaper the size, the more costly it is to remove.

For this reason a more expensive alternative is sometimes used — carboxymethyl cellulose (CMC) produced as we have seen (p. 8) by treating wood pulp or cotton linters with alkalis. Sizes and finishing agents are also made from natural plant gums, and include locust bean gum (from *Ceratonia siliqua*), and gum arabic (from *Acacia senegal*). Seaweeds too, provide us with size materials; *Macrocystis pyrifera* and *Laminaria* species are important sources of alginates used both for sizing and the thickening of printing inks in very large quantities.

Cassava tubers (Manihot esculenta) on Piro Indian cloth from the Peruvian Amazon. Starch from these tubers is used as a size

In Japan, a type of seaweed glue, funori, is widely used for sizing, stiffening and glazing cloth. The principal seaweed involved is *Gloiopeltis furcata* which is abundant in the warmer waters off the Japanese coast. It is gathered throughout the year with the help of long-handled hooks or rakes which detach it from the rocks on which it grows. This red seaweed is converted to size by simply dissolving it in hot water once it has been bleached and dried.

The best quality agar used for sizing is also made in Japan, from *Gelidium* seaweed.

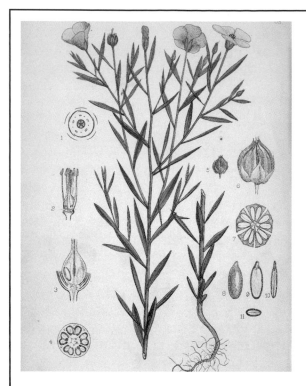

FLAX

Linum usitatissimum

The flax plant with its wiry stems and distinctive soft blue flowers is just one of some 200 species in the *Linum* genus. Used by people since pre-historic times, flax provides both fibres from its stem for linen, and linseed oil which is crushed from its seeds.

Different varieties (with different coloured flowers) have been developed to maximize production of these raw materials: those with long and relatively unbranched stems up to one metre high are grown for fibres, whilst the shorter-stemmed varieties with larger seeds are cultivated for oil.

The flax grown for use in textiles is harvested not by cutting, but by pulling the stems, to preserve the full length of the fibres which run along them. The flax 'straw' as it now becomes is then dried and the seed pods removed. In order to extract the fibres the straw must be 'retted', a process that breaks down the plant material that binds the fibres together. In France, where most of Europe's flax is grown, this is done by leaving the straw on the ground for 3 to 6 weeks so that the action of the dew will gradually rot away the non-fibrous parts. In Belgium flax straw is usually soaked in tanks of warm water, a process taking only 5 to 7 days. The next stage is known as scutching and it involves the retted straw being fed between cylinders which separate the fibres and remove any woody pieces or bark. In this way, both long and short fibres are obtained for textile use together with a certain amount of 'shive', the waste woody matter, which is sometimes made into panel materials, such as chipboard or plasterboard used in construction work.

After scutching, the flax fibres are combed and slightly twisted before being spun into yarn. Traditionally they were bleached in the sun to remove the natural yellow colour (which the expression 'flaxen-haired' refers to) but most flax is now bleached chemically.

To produce very fine and supple yarns the spinning process takes place under water, which softens the fibres. Renowned for their hard-wearing properties, the yarns are finally knitted or woven into cloth alone or mixed with other fibres such as wool and cotton.

The special processing required by flax fibres, unlike cotton, is very time-consuming and has never been highly mechanized. In addition, linen made from hand-processed flax is of better quality than that prepared mechanically. The current high price of linen reflects both these factors.

Linen

The main purpose of sizing cotton yarns is to give strength to the warp threads which will form the cloth's backbone. Until the early nineteenth century, when cotton ceased to be a luxury, these warp threads were generally replaced in the hand-woven cottons of Europe and America by the much stronger fibres of another plant: flax (*Linum usitatissimum*).

Flax was in fact the principal source of plant fibre for all the clothes worn on both these continents until the time of the Industrial Revolution. A slender, herbaceous plant with striking sky-blue flowers, flax is indigenous to western Europe and since prehistoric times has been its most important natural fibre. It is thought to have been cultivated at least as long ago as cotton, which developed in the warmer regions of the world. Unlike cotton, however, which is produced from seeds, flax fibres are taken from the stem of the plant for weaving into linen.

Fragments of different types of linen, as well as yarns and rope made from flax, have been found in the remnants of Swiss Lake Dwellings which date from about 8000 BC. The Egyptians wore fine linen clothing and used the cloth to wrap the bodies of their dead. Long before the Christian era linen manufacture had become part of domestic life around the Mediterranean basin and both the Greeks and the Romans appreciated its lustrous quality and texture. Some early linen fabrics were so fine, with up to 500 threads per inch, that pieces the size of an overcoat could be pulled through a ring the width of a £1 coin!

Whilst helping it compete with other fibres which are now processed more cheaply, the special qualities of linen have singled it out as a modern luxury material. Flax fibres are two to three times as strong as those of cotton and, up to one metre long, are naturally smooth and straight. Our English word 'line' is said to be in part derived from the Latin *linum* meaning flax, and this is now the plant's scientific name. The linear flax fibres have special water-absorbing and drying qualities too, enabling fabric made from them to 'breathe' well, and a lustre that is produced by their capacity to reflect light from both inner and outer surfaces.

Russia grows the most flax in the world today, but France, Belgium and Holland grow the most in Europe, with an annual harvest of some 450 000 tonnes a year. About forty European countries currently spin flax fibres into yarn for use as sewing thread, twine and webbing, as well as for weaving or knitting into material.

Despite the stone-age ancestry of flax in Switzerland, the Romans are the people credited with introducing the plant and its technology to France and Belgium. The Emperor Charlemagne established centres of linen weaving in several Flemish cities in the ninth century, and by the late Middle Ages much of Europe's woven cloth was made of a mixture of wool and linen. Most European monasteries cultivated flax, as linen was prescribed for ecclesiastical garments and shrouds.

Britain's naval strength developed with the aid of linen sails and ropes made from home-grown hemp (*Cannabis sativa*). Both these raw materials became so important that Henry VIII required everyone with 20 acres (8 hectares) of land suitable for cultivation to grow one or other of them on some part of it.

By the mid-eighteenth century, with the emigration of protestant linen workers from Roman Catholic Belgium and France, Ireland had become a major centre of linen weaving. Irish linen and fine handmade lace came to acquire world fame, a reputation that still exists

Irish linen

today. From Ireland, the skills of flax weaving were taken to North America with the early colonists, and a sturdy cloth 'linsey-woolsey' made from flax and wool — which was still the principal textile fabric in temperate areas — became the main clothing material in the colonies.

It was not until the nineteenth century that cotton grown cheaply in America and woven into cloth in England overtook linen as the major plant fibre used for making clothes.

Getting tough with plants

That most famous item of modern clothing — the pair of jeans — has come to be associated entirely with cotton. Available in any number of ever-changing styles and textures, perhaps the jeans you bought last summer had rips cut in them or were rhinestone studded!

When Mr Levi Strauss made the first pair of jeans, however, he intended them as tough protective working clothes. They were made not of cotton but of a much tougher fabric: hemp (*Cannabis sativa*). The cloth he used was imported from Nimes in France and the French name for this material 'serge de Nimes' was soon corrupted into 'denim'. Later the French pronunciation of the Italian city Genoa — from which the cloth was also exported — gave rise to the much used name 'jeans'.

Once cultivated in Britain and many other parts of Europe, hemp is hardly used today in western clothing. In fact, to grow the plant at all in Britain without a special license is a criminal offence! The source of marijuana, hemp is now much more famous for the resinous chemicals produced by all parts of the plant than for the original cause of its popularity; its stem fibres. Like those of flax, the fibres are very strong but are generally stiffer, since they contain more lignin. They are removed from the stems in a similar way, by retting and scutching and are spun into fabric that is somewhat less flexible than linen.

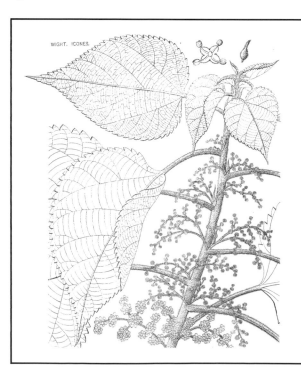

RAMIE

Boehmeria nivea

Although they lack the elasticity of wool or silk and the flexibility of cotton, ramie fibres can be separated almost to the fineness of silk and have been made into lace and virtually transparent fabrics in China.

NETTLES

Before cotton cloth became cheap enough for general use in Europe, the fibres of the common stinging nettle *Urtica dioica* were used as well as those of flax and hemp for making cloth.

In Northern France and Germany especially, where it was known as 'nesseltuch' these nettle fibres which were extracted from the plant's stem could be bleached to make a cloth as white as linen.

Ramie (*Boehmeria nivea*)

Both cloth and rope made from hemp were used in China and the East over 4500 years ago, and have been widely used in the past in southern and central Europe, Japan and India. During the eighteenth century cultivation was mandatory in the American colonies and in England for the production of rope and materials used for caulking the hulls of clipper ships. Canvas too, was traditionally made from cannabis, from which it takes its name.

The main hemp-growing countries today are, in order of production: the USSR, Italy, Yugoslavia, Hungary, India and China.

Are you wearing Chinese nettles?

Another very tough fibre—in fact eight times stronger than cotton and three times stronger than hemp—is ramie. Looking rather like the stinging nettle to which it is related, but without the sting, two types of *Boehmeria* provide fibres that are known as ramie. Many native names exist for the plants, but ramie proper (*Boehmeria nivea* var. *tenacissima*) is generally called green ramie or rhea, whilst the better known *Boehmeria nivea*, with soft white hairs on the underside of the leaves is called white ramie or China grass.

White ramie is a hardy perennial shrub which can grow to a height of 2 metres. About 70 per cent of the plant consists of fibres which are very tough and water-resistant; in fact, their strength is increased when wet. To extract them from the bark and sticky outer tissues in which they are contained (the phloem), these parts are first peeled from the stems and then scraped and washed or chemically treated. Processing is traditionally done by hand, but in the more industrialized countries machines do most of the work and caustic soda is used to free the fibres from the gums and pectins which stick to them. Before spinning into yarn the fibres are often chemically bleached and also softened, to stop them from becoming too brittle.

After perfect processing, ramie fibres are said to be the longest, strongest, silkiest and most durable of all those extracted for human use from plants. Because of the difficulties of harvesting, however (the plants are perennial and tend to grow in uneven stands), and of extracting the large amounts of gums and pectins in the stems, ramie has not been grown as a major Western fibre crop in the past. Yet the outstanding qualities of ramie have long been appreciated by the peoples of southern China and northern India to which white and green ramie respectively, are native. In some parts of Asia the plants have been cultivated for over 7000 years, and the ancient Egyptians were also familiar with them. Ramie plants have now been introduced to the warmer parts of Europe and North and South America, and Brazil has become the main exporter. China and the Philippines, both traditional producers for their own markets, also export the fibre.

In the last year or so you may have bought a heavy sweater made of Chinese or Philippine ramie mixed with wool or cotton. Though British and American trading laws forbid the importation of foreign man-made fibres, ramie as a natural fibre has been imported in large quantities and made, amongst other things, into winter knitwear! Other more traditional uses of ramie include fishing nets, cordage, sacking and carpet backing, upholstery materials, sailcloth, table linen and sheeting.

Ramie fibres may be woven into cloth and lace

Mulberry trees for royal clothes

Perhaps the best known wedding dress in history — The Princess of Wales' — seen simultaneously by several hundred million people on television sets around the world, was made in part by caterpillars! Thousands of silkworms, fed on a diet of mulberry leaves, spun 180 metres of silk to make the hand-embroidered dress, which was adorned with antique lace.

Though silk, the most expensive fibre in the world, is not produced directly by plants, it could not be made without them. The most famous are the mulberry trees, whose leaves provide the staple food for the common silkworm *Bombyx mori*. Hatched from the eggs of silkmoths, these white caterpillars eat the leaves until they start to spin their silk cocoons which house them as they transform into moths. Most silkworms will never reach this adult stage however, as their cocoons are carefully unravelled and made into fine silk thread.

Two species of the very large mulberry family, *Morus nigra* and *M. alba* have become the traditional food of *Bombyx mori*, which produces the finest white to yellow silk and also the largest proportion of the world's silk supply. Both are believed to be native to Asia; the white mulberry *M. alba* to the mountainous regions of Central and Eastern China and the black or common mulberry *M. nigra* to the mountains of Nepal and the southern Caucasus.

Whilst they have been grown for their edible black fruit, and in the case of *Morus alba* for its bark which was a traditional material for Chinese paper-making, their use as food for silkworms stretches back thousands of years. The Chinese had developed the art of sericulture

Black mulberry (Morus nigra)

White mulberry (Morus alba)

MULBERRY

Morus nigra and *M. alba* are deciduous trees growing to a height of 6 to 9 metres. The leaves, which cover the densely spreading branches, contain traces of essential oils that are attractive to silkworms and a sticky, rubbery sap that is believed to influence the composition of the silk. Each strand of silk, which is produced by glands and secreted from the silkworm's mouth, is three times stronger than a strand of steel of the same density!

Larva of Bombyx mori, *the common silk worm*

at least 5000 years ago, and it was they, along with the Persians, who supplied the world with silk for many centuries, exporting it across Asia in famous silk caravans. Though the art of silk-making was kept a closely guarded secret, the mulberry trees themselves were gradually taken westwards from Persia to Greece and other parts of Europe, apparently reaching Britain with the Romans.

The long-lived mulberry trees became popular in the Middle Ages and were prized for their blackberry-like fruits which added colour and flavour to wine or were eaten as a purée. Mulberry trees were often grown in medieval gardens. By the eleventh century, however, the secret of silk-making itself had spread via India and Arabia to Sicily and southern Spain. In the sixteenth century Elizabeth I encouraged the commercial planting of mulberry trees at Rye in Sussex and in 1606 James I imported one million black mulberry trees, hardier than *Morus alba*, from France. The idea was to start a silk industry in Britain, but the bitter complaints of silk merchants who feared the loss of their trade, plus feeding difficulties with the silkworms caused the project to collapse.

Silk has long been synonymous with luxury. The Romans were quick to appreciate its sensuous qualities, but the Emperor Tiberius prohibited its use by men, believing it to be effeminate. This historic ruling, however, did nothing to discourage King Henry VIII from wearing the first pair of silk stockings to be seen in England!

Despite the abandonment of silk stockings by men — and also more recently (in the face of competition from synthetics) by most women — and the general availability of the material in other forms today, silk continues to be linked with royalty. The silk for the Princess of Wales' wedding dress was spun by silkworms raised at Britain's only silkfarm near Sherborne in Dorset. Between three and four million silkworms are reared there each year, principally on mulberry leaves, which they consume by the tonne. But there are many different sorts of silk-producing caterpillars world-wide, which feed on different kinds of plants.

Bombix mori, the 'true' silkworm, which has been domesticated in China for so long that the moths can no longer fly, eats only mulberry. In Japan, which was the major silk-exporting country for several centuries, some 700 varieties of mulberry are known.

Tussah silk, on the other hand, is spun in India and Pakistan from the cocoons of wild silk-producing larvae, which feed on sal leaves (*Shorea* spp.) and a hybrid species also reared in quantity in China, feed principally on oak. Another sort of silk, 'Eri', which is creamy white or reddish in colour, is produced by larvae that feed on castor-oil plants.

Silk sari from India *Japanese women in silk kimonos*

China and Japan produce three-quarters of all the silk in the world today, but India, Italy and Korea export very large quantities of silk fabrics and yarns. The breeding of silkworms and other silk-producing larvae is a largely rural and very labour-intensive process involving around 2.5 million farm workers in 30 countries world-wide. In India most of the silk is made into saris, whilst the kimono accounts for 90 per cent of all the silk used in Japan. In fact, at least 25 per cent of all the world's silk goes into Japan's traditional dress, but the amount is falling as these costumes are being abandoned and the land reserved for mulberry cultivation is used for other industries.

A PLANT THAT SAVES LIVES

Some of the lightest and smoothest of plant fibres may have helped to keep you dry or even saved your life. The silky hairs that surround the seeds of the kapok tree (*Ceiba pentandra*) are too fine and slippery to spin but are used instead mainly as a stuffing material. Life-jackets and buoys have traditionally been filled with kapok since the cells of the fibre are full of air whilst the cell walls are impervious to both air and water making them naturally buoyant and light. Kapok fibres have excellent thermal insulation properties and are used to stuff outdoor jackets and sleeping bags, as well as cushions and mattresses. Their resistance to attack by insects also makes them very suitable for these uses.

Designed to help disperse the seeds to which they are attached, kapok floss or fibres develop inside large cylindrical pods which split open when they are fully ripe. For commercial use the pods are gathered just before this happens and the floss, which falls easily from them once they are open, is spread out on the floor of large open-sided sheds to dry. Frequent turning ensures that the floss dries evenly, gradually expanding as it does so before the seeds and other foreign matter are removed. Kapok fibres are so light and fluffy that workers must wear fine protective masks and the drying sheds themselves are enclosed in fine wire mesh to stop the fibres from floating away.

Kapok comes from one of the tallest trees in the tropical forest, which has spread from South and Central America to various parts of Africa, the Philippines, Sri Lanka and Indonesia.

Before the arrival of the Conquistadores, Mayan and Aztec peoples revered the tree and regarded it as sacred. Its size and stature led them to regard it symbolically as a link between

THE PINEAPPLE SHIRT

If you thought silk and cotton made the finest shirts, have a look at one made of pineapple fibres!

The Philippine Islands have a long history of making cloth as fine as gossamer from fibres extracted from the leaves of the pineapple (known there as piña) and this highly skilled practice continues today. The industry, which has its centre in Aklan on the island of Panay, is still largely unmechanized and most of the piña fibre is woven by hand. Different pineapple varieties can be used, but the one with the highest fibre content and which also gives the finest fibres is known as Red Spanish or Philippine Red (*Ananas comosus* cvs.). Young fruit are pinched from the stem before they can develop thus channelling all nourishment to the growing leaves. To prepare the fibre the leaves, which are cut from the pineapple plant at intervals, are first scraped by hand with the aid of a knife or broken porcelain plate or saucer. This is a very time-consuming process requiring much patience. Having scraped the leaves, the fibre bundles which are now revealed are cleaned by further scraping with a clam shell under running water and are then dried in the sun. Beating with a bamboo stick separates the fibres and dislodges any other material that may still be attached. The end of each fibre is knotted individually and a yarn ready for weaving on a hand loom is formed by gluing several fibres together.

In an attempt to update and mechanize the production of piña cloth, the Philippine Textile Research Institute has promoted the development of new techniques and modern machinery to replace the time-consuming preparation of the fibre by hand. The separation of the fibre bundles from the surrounding tissue by microbiological

retting, chemical degumming in dilute solutions of sodium hydroxide, and mechanized processes for softening, carding, spinning and then weaving or knitting the fine, lustrous fibres into cloth are all underway.

To weave a metre of piña cloth by hand takes a fast worker about a day. About 60 fresh pineapple leaves will be needed for this amount of fabric, yielding some 24 grams of piña fibre. Traditional items of clothing made from piña cloth — said to be more delicate in texture than that made from any other vegetable fibre — include fine shawls and the 'Maria Clara' and 'terno' costumes for women, as well as 'panuelos', scarf-like fabrics stiffened with starch, and shirts for men. The Philippine national costume also comprises garments made from piña fibre.

Kapok pods

the earth and the universe, its deep root system and magnificent spreading canopy seeming to support the sky whilst reflecting the social ordering on earth.

The kapok tree's longevity and strength were also qualities that the Mesoamerican rulers much admired. Since they were seen as the providers of daily sustenance for their people — a rather precarious position maintained with the help of elaborate rituals believed to determine agricultural success — the native rulers used symbols of this majestic tree, important for its shade, fruit and seeds, to represent their own power and ability to support the populace. Today, Indonesia, Thailand and India are the world's largest producers of kapok, and between them they export thousands of tonnes of the fibre each year.

LINGERIE FROM LOGS

Some of your slinkiest and silkiest clothes, and some of the most hard-wearing, will almost certainly be made from wood pulp! Everything from underwear and swimwear to flame-retardant military combat uniforms, could all be made from eucalyptus logs!

Synthetic or man-made fibres now account for almost half the total fibre consumption in the world. Whilst many of the most familiar such as nylon, acrylic, terylene and polyester are derived from oil (itself the result of the compression of plants for millions of years), a number are made directly from the cellulose contained in living plants. Viscose (a 'rayon' fibre, the name given to the first artificial silk), acetate and tri-acetate, are perhaps the best known of these.

It was the desire to produce a material with the look and texture of silk, but not the cost, that brought about the invention of the first synthetic fibres. As early as 1664 Dr Robert Hook had thought that there might be a way to make 'an artificial glutinous composition' to rival the fibre produced by the silkworm, but it was not until the mid-nineteenth century that the first discovery of the usefulness of plant cellulose was made — not just for fibres, but for materials that would later be used to make hard films and plastics.

In 1846 a German chemist Christian Schonbein made the crucial discovery that cellulose in cotton could be made soluble. This happened when he used a cotton apron to mop up some sulphuric acid and saltpetre he had spilled. As the substances mixed, the apron was inadvertently turned into cellulose nitrate and exploded as it dried! The scientific interest that the event initially aroused was directed, not surprisingly, at the potential of this new but unstable compound an an explosive. But more inventions followed. In 1885 Sir Joseph Swan produced cellulose nitrate in the form of filaments and yarn which he used as electric light bulb elements.

Some seven years later, three British scientists discovered and patented another process for dissolving cellulose by converting it into sodium cellulose xanthate, or viscose. A little later they discovered a way of making cellulose acetate — another soluble cellulose derivative that could be spun into fibres or made into solid plastic.

Schonbein's apron had reacted so spectacularly when it soaked up the acid because cotton fibres are almost pure cellulose. Cotton linters (the shorter fibres produced by cotton seeds and otherwise unusable for textiles) thus came to supply the raw material for the new fibre industry, but as demand out-stripped supply, woodpulp was substituted and remains the form in which most cellulose is commercially available today.

The type and source of timber made into pulp has changed considerably in recent years. Traditionally the pulp used in Britain was made mostly from softwoods such as spruce and pine species from northern Europe and Scandinavia, with the addition of birch, beech and aspen and other common temperate hardwoods. But the demand for ever-increasing amounts

WOOD FIBRES

Wood fibres, the walls of which are mainly made of cellulose, are so strong and tough that they can only be separated with the help of powerful machinery and chemicals.

Once felled, logs are stored for several weeks to reach a uniform moisture level before being cut into small chips ready for pulping. They are then fed into computer-controlled digesters where various chemicals dissolve the pectins and liberate the fibres in the wood. This 'cooking' process reduces the logs to cellulose pulp. The pulp is now washed and bleached and finally pressed into thick sheets ready for making into viscose or other synthetic materials.

Making viscose

To produce viscose fibres, woodpulp which has been compressed into thick sheets like blotting paper is first steeped in a solution of caustic soda. This converts the pulp to alkali cellulose. After pressing to remove excess liquid the resulting material is ground into fine crumbs and added to another chemical, carbon disulphide which reacts with it to form sodium cellulose xanthate. More caustic soda is used to dissolve this compound and it is then carefully filtered before being squeezed out through the fine holes of a spinneret or jet, into a bath containing a mixture of salts and sulphuric acid. This final stage solidifies the new viscose fibres and they are now either spun into yarns directly in the form of continuous filaments — principally for dress-making fabrics and furnishings — or cut into shorter lengths for spinning (like cotton fibres) into other sorts of viscose cloth.

Acetate and tri-acetate fibres (such as 'Dicel' or 'Tricel') which are finer and more glossy then viscose, and widely used for finer softer fabrics and linings, are produced in a similar way, but the wood pulp is mixed initially with acetic acid before processing with acetic anhydride.

Outfit (left) made entirely of viscose, derived from wood chips (top)

of pulp brought about the need for faster-growing trees, and tropical and subtropical woods such as eucalyptus and acacia now form a substantial part of the dissolving pulps produced.

Much of the pulp used in Britain today comes from huge plantations in South Africa and Brazil. Both these countries have encouraged major investment in eucalyptus trees which will produce usable timber in 8 to 12 years from seedlings grown in nurseries. Certain species will also produce shoots that can grow up to five metres per year after the trees have been felled, and careful coppicing is practised in some areas to ensure a continuous supply of wood.

TREES FOR BABIES

Woodpulp keeps millions of babies comfortable and dry. It is the basic raw material for almost all disposable nappies and their soft, absorbent padding. In Britain alone we use over 2.5 billion disposable nappies every year, spending around £330 million in the process!

The trees involved will vary according to the country in which the nappies are made and its source of supply. Most nappies available in Britain and the rest of Europe are made from Scandinavian trees — the majority softwoods — which are reduced to pulp by chemical processing. Only 45 per cent of each tree is suitable for this kind of processing, the rest being burned as waste.

The pulp used to make the fluffy padding which forms 70 per cent of the nappy has traditionally been bleached using chlorine-based chemicals to give it a brilliant whiteness. One major Swedish nappy producer, however, is now using a mostly mechanical process that is totally chlorine free, to make the nappy padding. This has two major advantages: no harmful chlorinated waste is washed into our rivers and seas (as would be the case otherwise) and more than 90 per cent of each tree can be used. By this new method, one fully grown pine tree now makes enough 'fluff pulp' to fill 1000 disposable nappies, doubling the amount made by conventional means.

Thousand of acres in Natal province and Zululand are currently planted with *Eucalyptus grandis*, a very fast growing species introduced from Australia and well-suited to local climate and soil conditions. Other imported species such as *E. paniculata* and *E. fastigata* have been grown at different times in South Africa, along with the black wattle (*Acacia mearnsii*) which now forms up to a quarter of some South African wood pulps. Another supplier of woodpulp made from eucalyptus — this time *E. globulus* — is Spain. But the world's biggest factory producing 'rayon grade' dissolving pulp is situated in South Africa, near Durban. It supplies about 450 000 tonnes of dissolving pulp a year, not just to the fibre industry, but to the makers of paper, plastics, cellophane and carboxymethyl cellulose.

Generally speaking, the tree species used to make pulp will depend both on where the mills are situated and the economics of growing — or importing — plantation trees. The USA, which produces most of the world's total dissolving pulp uses a variety of mixed 'southern hardwoods', including native oaks and maples as well as the familiar softwoods like western hemlock and various pines, which are also grown in large numbers in Canada. India, on the other hand, uses a mixture of tropical hardwoods and bamboo, whilst Japan and Thailand — themselves consumers of enormous quantities of pulp — use mangrove trees which grow abundantly in coastal regions of the tropics to supplement yields of non-native eucalyptus.

Despite the prevalence of eucalyptus, many European countries import significant amounts of tropical hardwood pulp if the price is right, often cut from remote areas of virgin rainforest where clear-felling continues unabated. In the face of the continuing destruction of these fragile and irreplaceable forests, the ecological advantages of man-made plantations are obvious.

BLACK WATTLE

Acacia mearnsii

The province of Natal in South Africa has been the main black wattle producing area for over half a century. It was originally introduced from Australia as a shelter tree to plant round crops and homesteads, but plantation cultivation was encouraged to provide a replacement for depleted South American quebracho trees which had come to supply most of the world with tannin extracts.

Today this fast-growing tree supplies much of the world's woodpulp too, for use in the paper, textiles and plastics industries. Most of the work involved in felling and handling the trees is done by Zulu men.

Spectacle frames

One of the end products that can be made from black wattle pulp after processing with acetic anhydride is cellulose acetate. Spectacle frames, toothbrushes, combs, buttons, knitting needles, pens and car steering wheels are just some of the more familiar items made from moulded plant cellulose.

DYEING AND PRINTING WITH PLANTS

Most of our clothes are dyed or printed. For thousands of years plants have supplied most of the basic raw materials for these operations and in many rural societies throughout the world they continue to do so. Natural dyes in particular still play an important role in developing countries, where local craftsmen do not have access to synthetic colourants.

Nearer home, herbal dyes whose colours are acknowledged to be richer and more subtle than chemical equivalents have continued to arouse much interest, but they tend to be used only by dedicated individuals and those who work in small cottage industries.

Dyeing in industrialized countries today is based entirely on cheap standardized aniline dyes derived from coal tar. If you were to buy a piece of modern clothing coloured with vegetable dyes the fact would certainly be well advertised. What is never obvious, however, is the extent to which plants are still used as vital auxilliaries to the dyeing and printing processes. Derivatives of commercial plant oils for example, such as coconut, palm, soyabean and linseed play an important part in the manufacture of detergents used for cleaning crude wool before it can be spun. These oils also help prepare and soften the surfaces of many woven and knitted fabrics before printing or dyeing can take place. Pine oil from a number of pine species grown in Europe, America and India is widely used too as a wetting and dispersing agent, helping to ensure that dyes will 'take' properly and spread evenly when applied.

Plant gums are particularly useful. Carrageenan made from seaweed, and gum arabic, the reddish sap tapped from spiny acacia trees that grow wild in Africa, are used to give fabrics a soft finish and compact feel, as well as an even surface for printing. Gum arabic was in fact vital for the invention of the lithographic printing process in the eighteenth century. Indispensable for this technique today, the application of gum to the parts of the plate required to print an image makes them receptive to the printing ink, while the areas left free will repel it.

An even pick-up and consistency of dye or ink, vital in large commercial dying or printing operations, is also ensured by the use of natural gums, but their other main use is as thickening agents. Gum karaya, gum tragacanth, guar and locust bean gums are all used for this purpose. They control the flow and elasticity of textile dyes and inks and allow multiple patterns to be printed that are sharp and bright, without penetrating the cloth too deeply.

The solubility of these gums and their derivatives, plus their resistance to alteration by other chemicals, makes them highly valued in the textile trade.

Other ink thickeners include alginates and natural plant starches such as potato, maize, cassava and wheat. Very large quantities of all these natural materials are used today.

Woad (Isatis tinctoria)

DYE PLANTS IN EUROPE

By the Middle Ages, dye plants had become so important that large areas of agricultural land in Britain and Europe were used to cultivate them. Woad for blue (*Isatis tinctoria*), weld for yellow (*Reseda luteola*) and madder for red (*Rubia tinctorum*) were the most important.

Later, dyes from tropical trees were imported in vast quantities, brazilwood (*Caesalpinia* spp.) for red and orange colours and logwood (*Haematoxylum campechianum*) for black. Logwood was used for dyeing silk in Mexico until the 1960s.

With the invention of the first synthetic dye in 1856, a lavender colour made from coal tar, the use of plants declined rapidly. In the Scottish Hebrides, only a few weavers still colour the tweed materials they make with plant and other natural dyes.

A number of very familiar temperate plants will produce delicate dyes when 'fixed' with chemical mordants. Onion skins, ragwort, elder, privet leaves and heather are just a few of those that yield a range of yellows. Green can be obtained from bracken, ivy and stinging nettles amongst others, whilst reds and browns are produced from a number of lichen species, privet berries, blackberries, cherrywood, dandelion, dock and lady's bedstraw. Almost all the textiles made before the mid-nineteenth century from all over the world, exhibited in museums today, are dyed with plants and natural minerals.

L'Anil ou l'Indigo
Indigofera tinctoria Linn. Sp. Pl.

Indigo

Blue is one of the rarer dye colours to be produced by plants. Though Europe had its own source in woad (used to colour the clothes of ancient Britons as well as the outfits worn by Robin Hood and his merry men) it could never really equal indigo for richness. Several plants of the genus *Indigofera*, which are legumes native to India, as well as *Lonchocarpus cyanescens* in Africa, produce the dye. Indigo dye was already an important item of Indian trade several thousand years ago and became famous for its rich and beautiful colour. The process of producing it is complicated and varies from place to place. In Nigeria this involves mixing the leaves of indigo plants — which are pounded and then formed into balls to ferment as they dry — with solutions of wood ash.

The strong blue colour is produced only after cloth steeped in this solution is exposed to the air, thus activating the latent colouring compounds.

Wool dyed with iris roots

PLANTS ON THE SOLES OF YOUR SHOES

When it comes to shoes, nothing quite matches the quality and feel of leather. Chinese warriors, Egyptian nobles and England's Anglo-Saxon men and women all used leather to protect their feet.

Most of us own leather shoes or boots, and take this strong and versatile material for granted. Without the grasses, grains and forage crops which feed our animals, there would of course be no leather, but it is the compounds in a range of other plants that have traditionally given it its special qualities. These compounds are known as tannins and until the end of the last century when chemical alternatives were introduced, all hides were tanned with them to turn them into leather. Untanned hides exposed to moisture would very soon disintegrate and rot. Our museums contain several thousand examples of Roman and Medieval leather shoes and sandals, excavated by archaeologists in a perfectly preserved state only because they were tanned with plant materials.

Schinopsis quebracho-colorado

QUEBRACHO

Schinopsis quebracho-colorado, S. balansae

Several different Latin American trees with very hard wood have been called Quebracho in the past, but only two *Schinopsis* species, *S. quebracho-colorado* and *S. balansae*, have been exploited on a commercial scale for the production of tannin extracts.

The slow-growing, very dense wood of these trees which take 80 years to reach maturity is so hard that it frequently broke the axes and the teeth of circular saws used to cut it down. But this did not stop the trees supplying more tannin for world use for many years than any other vegetable material. The ruthless over-exploitation of the trees, which grow singly or in small groups over some 520 000 square kilometres of the Argentine and Paraguayan Chaco, led to a very serious depletion of numbers.

The heavy heartwood of quebracho trees — which turns a dark reddish brown after exposure to the air — is cut into chips and soaked in special vats to extract the tannin. This extract is particularly suitable for the sole leather of shoes, as it has weight-giving properties and produces a tough, firm feel. Eighty thousand tonnes of quebracho tannin extract were exported from South America last year.

A wide range of plants contain tannin, but certain trees which have a high content have come to be most widely used. Traditionally, countries used the sources that were plentiful and close to hand. England, for example, depended very much on oak bark, but today those countries that are most industrialized tend to import tannin, often in spray-dried extract form, from the largest and cheapest commercial sources.

Because they must withstand great heat as part of the shoe making and moulding process without shrinking, and also be very flexible, the leather uppers of almost all the shoes produced in bulk today are tanned with chromium. Most soft and pliant leather clothes and gloves are also made in this way. But leather shoe soles and accessories that need to be tougher, more resilient and almost stretch free — handbags and belts for example — are tanned with plant extracts. Most leather that must stand up to heavy industrial use, as well as traditional items such as saddles, are also tanned with vegetable materials.

The current imported favourite in the USA, Japan and Britain is mimosa or wattle bark which comes from the Australian tree *Acacia mollissima*. In 1986 these three countries imported nearly 18 000 tonnes of extract. The black wattle as it is more commonly known is grown in huge plantations in South Africa, Kenya, Tanzania and Zimbabwe, and most recently Brazil.

The next biggest tanning extract import is quebracho, the name given to a number of related trees of the genus *Schinopsis*, indigenous to the Chaco region of Paraguay and Argentina. The very tough quebracho wood (which has taken its name from the Spanish 'quiebra hacha',

meaning 'it breaks the axe'), has supplied most of the world trade throughout the present century. The wild stands of these very slow-growing trees became so depleted, however, that plantations of fast-growing black wattle were established in South Africa to replace it.

The third favourite of the UK, USA and Japan is sweet chestnut (*Castanea sativa*), of which nearly two and a half thousand tonnes were imported in 1986. Chestnut is a traditional source of tannin for many European countries, and up to the 1950s another variety, *C. dentata*, was grown extensively in the USA.

France and Italy are two of the major suppliers of sweet chestnut today. With its traditional prowess in the fashion trade and in the face of a sharp decline in consumption elsewhere, Italy is now Europe's largest producer of vegetable-tanned leather for shoes.

Other traditional supplies of tannin have come from the acorn cups of the valonea oak (*Quercus aegilops*) which is native to Turkey, and the sumac tree (*Rhus* spp.) from the same area. Observing the practices of indigenous Indians, North America's colonizers used the bark of native hemlock trees (*Tsuga* spp.) for tanning leather. Australia meanwhile has always had a plentiful supply of wattle and different eucalypts, some of which have very high tannin contents.

In tropical regions mangrove bark is widely used and cutch from *Acacia catechu*, which is indigenous to India and Burma is still popular. Another source from India is the fruit of the myrabolans tree (*Terminalia chebula*).

In industrialized countries tanning processes which were once very lengthy and laborious are now highly mechanized and very quick, lasting sometimes just a few days. After chemical cleaning and the removal of hair, hides are placed in revolving wooden drums containing tanning solutions. Different grades of 'tannages' as they are called are recognized, and since each extract has its own specific properties, different mixtures will be used for different ends. The tannins cause the collagen fibres in the hide to stabilize and become robust and durable, protecting them from microbiological decay. For some leathers, combinations of chromium and vegetable extracts are used.

After tanning is completed the leather is usually bleached, rinsed and then dressed with special chemical solutions. Plant oils such as rape seed, or gums like carrageenan are then used to smooth and polish the leather, helping to waterproof it and prevent it from cracking.

More plants for our feet

The prevalence of synthetic compounds such as polyurethane and PVC for making shoe soles today has meant that the amount of leather tanned with plant extracts has decreased sharply. But other plant materials are still to be found in our footwear. Hardwearing fabrics such as canvas made from natural fibres, may form the uppers, whilst synthetic fibres made from woodpulp as well as cotton may be tying up our shoes.

Natural rubber is in evidence too. Many wellingtons and agricultural boots are made from a base of natural latex and it is used as an adhesive especially for linings in many modern shoes.

Many modern wellington boots are made from a base of 40 per cent coagulated natural rubber

A shine from plants

Whether our shoes are made from leather or PVC it is likely that we will polish them with waxes made from plants. Since it is harder than beeswax, carnauba wax from the Brazilian wax palm is one of the main plant waxes used, forming some 10 to 15 per cent of the total volume of some products. Because of its hardness only small quantities are needed to produce a shine.

Candellila wax from Mexico is also found in some shoe polishes and wax from esparto grass and bamboo species has been used in different regions. Carrageenan from the seaweed known as Irish moss may be present in polishes too, as it holds down and smoothes out the tiny rough projections on the surface of the leather.

AND TO CAP IT ALL!

Some people say that no outfit is complete without a hat! Britain's Royal Ascot and many a genteel garden party would not be quite the same without this elegant accessory. Traditional boaters, sunhats, and many creative designs, especially for women, are still made from a base of different plant materials, but all of them are described by members of the hat trade as 'straw'.

In botanical terms, straw refers to the stalk on which a grain has grown or to a quantity of the dried stalks left after harvesting, but other commonly used plant materials including various palm and possibly *Agave* fibres all share this same general term, as well as man-made viscose fibres.

Today, almost all the good quality hat-making 'straws' used in Europe are either grown in or exported from China. The names they are traded under, such as 'xian' or 'paribuntal', tend to obscure further the identity of the plants involved and their precise identity is difficult to establish.

Hats and head-dresses for men and women have been worn as a means of adornment and as an integral part of costume since the earliest times. It is interesting to reflect that practically all the basic hat silhouettes of modern times were worn by ancient Mesopotamians — the

people of Babylon, Assyria and Chaldea — as well as by the Medes and the Persians, between the ninth and fourteenth centuries BC.

In the twelfth century AD, the fashions created in Paris, Venice and Florence had already become trend setters for the rest of Europe and the fame of the hats being made in Milan during the eighteenth century eventually gave rise to our term 'milliner'.

Woven straw hats as we know them in Britain date from the middle of the sixteenth century when Mary Queen of Scots brought straw plaiters back with her from France and a new industry arose. Plaiting, braiding and weaving of straw for hats was done in the home and became a source of income for women until the nineteenth century, when machines began to take over the intricate handwork.

The most valuable straw at that time — a wheat straw, which was cultivated especially for the purpose and cut whilst still green then bleached to a pale yellow — came from Tuscany. It was known as Leghorn after the hat-making city of Livorno through which it was exported and the large, finely plaited hats of the same name. Today, wheat straw is still used for some traditional hats such as the boater, with its distinctive flat crown. This hat shape was

Pith helmets were made from the pith of leguminous swamp plants native to India, covered with cotton cloth

popularized in the eighteenth century by Lord Nelson who incorporated it into the British naval uniform. Sailors stiffened and glazed the straw of their hats with varnish to protect them from the damaging sea-spray.

Though many of our modern hats are largely decorative, others, like the boater, were originally designed to protect the wearer from the elements or rigours of the climate. Two very well-known natural fibre hats, made famous during the last century, were designed to combat the fierce heat of the tropical sun.

The pith helmet or topee, which protected the heads of British army officers in India from the 1860s onwards, was made from the pith of leguminous swamp plants native to India; *Aeschynomene aspera* and *A. indica*. Covered with a white cotton cloth on the outside and lined with a green material, the inner cork-like pith of these plants (also known colloquially as sola or shola) formed the famous lightweight helmet which not only insulated the wearer against the heat of the sun, but was impervious to air and water.

The panama is another hat designed to protect the wearer from the powerful tropical sun. Unlike the pith helmet the panama is still widely used today. Famous, but misnamed, the hats come in fact from Ecuador and gained their name only because they were bought in Panama by the 'Forty Niners' on their way to California at the time of the Gold Rush. During the 1880s Ecuador began to export the hats in large numbers through Panama to the United States.

Today the small but thriving cottage industry centres around the southern Ecuadorian town of Cuenca. The fibre used comes from the leaves of a palm-like plant *Carludovica palmata* known locally as toquilla. The leaves are brought to Cuenca by the truck-load from the region of Montecristo and Jipijapa on the Ecuadorian coast. Since the hats are made here too, they are sometimes known as Jipijapas or Sombreros de Montecristo.

The toquilla plant grows to about two metres in height and produces large, fan-shaped leaves about one metre in diameter. These are harvested in succession from the base of the plant, while they are still folded. About fifteen are needed to make one panama hat. After drying, the fibres are separated from the leaves into strands with the thumb and fingernail or with a special comb, and may be further softened and bleached before weaving.

The best and most expensive hats, which may take several weeks to make, are so soft and finely woven that they can be rolled up easily like napkins and are commonly sold in narrow rectangular boxes made of local balsa wood.

Boaters outside Luton Town Hall, c. 1900. Luton is still a major centre of hat-making in Britain

From first foods to fast foods

Plants were almost certainly our first foods. Before people had developed their skills as hunters, herders and agriculturists, and many thousands of years before the advent of the pre-wrapped, pre-sliced loaf of bread baked from engineered strains of superwheat, native plants were our most immediate source of nourishment.

Around the world people experimented with and ate different parts of the plants they found growing locally. Many of these have been subject over long periods of time to continuous processes of selection and domestication, and as people have migrated, or discovered 'new' lands, so the plants have moved too.

About 12 000 of the estimated 250 000 flowering plant species in existence (not including the algae or the fungi) have been used by people as food, but only 150 or so have been cultivated to any extent. Today, twenty species alone provide 90 per cent of the world's food needs! Just two plant families contain some of the most important of these. The Gramineae or grass family gives us wheat, rice and maize which between them supply over half the world's calories, while the Leguminosae provides peas, beans and all kinds of pulse grains.

Other families important for the food plants they contain include the Rosaceae for fruits such as apples, pears, plums, cherries, almonds and apricots; the Cruciferae for many of our green vegetables; the Palmae for coconuts, dates, sago, oil and other palm products; the Euphorbiaceae for cassava; the Musaceae for bananas and the Solanaceae for potatoes, red peppers and tomatoes.

An increasingly large proportion of the major crops that dominate global food markets and which go to feed the people of the developed world are now the result of careful cross-breeding and genetic manipulation involving processes such as irradiation, chromosome doubling and tissue culture developed as part of the green revolution over the last fifty years. These crops are largely grown as monocultures, which means that each individual plant is genetically identical to its neighbour.

Though these techniques, along with modern farming methods have produced bumper crops and generally made harvesting easier, they have also facilitated the spread of virulent pests and diseases to which wild or less manipulated populations often have some natural resistance. Genetic material from wild plants continues to play a vital role in boosting our new engineered strains with characteristics that will help them thrive. It is for this reason that safeguarding the wild relatives of our crop plants — often found in regions where the native flora is now threatened by human activity — as well as the great fund of other plant species of both known and potential use to man, has become a prime concern of botanists around the world.

Unlike the people of many small, non-industrialized societies, who often grow a large number of crops in family plots or gardens and gather themselves almost everything they consume, the origin of much of what we eat and drink is entirely unknown to us. Whether eating a tomato or a bar of chocolate, we tend to worry more about its price than where it came from or, in the case of the chocolate, how it was made. Even where a 'country of origin' is stated, the fact that the size, colour and flavour of many fruits and vegetables are due to the genes of some little known relative, existing in a tiny scrap of forest on the other side of the world, is never apparent and certainly not advertised.

In the case of the chocolate bar and hundreds of other processed foods, the task of decoding and unravelling the list of ingredients often obscures any connection with plants at all. But plant materials are present in almost everything we are likely to eat or drink, from fish fingers to champagne. Where they do not form the basis of the food itself, colourings, stabilizers, thickeners and sweeteners are some of the commonest forms in which they will appear.

Without plants of course — and particularly the grasses — there would be no milk, butter, cheese or yoghurt, and no meat. Around 90 per cent of the grain harvest in the USA goes to feed cattle, and most livestock, whether factory-farmed or free-range, eat mixtures of protein-rich cereals and plant fodder in various combinations.

Many of our foods are packaged in plant-derived materials too, generally wood pulp, which is transformed by mixing with chemicals into paper, cardboard, cellophane and other protective wrappings. So as you unpack your carrier bag, spare a thought for all the different plants you have just bought!

PLANTS FOR BREAKFAST

The chances are that whatever we decide to eat for breakfast, most of it will have come directly from plants. Here we start by looking at a typical continental breakfast, which offers us a selection of foods and drinks we are likely to consume in other meals and snacks throughout the day, and at the plants that produce them.

Bread

Available in an endless variety of shapes and forms, bread is indispensable to most European breakfasts. The British certainly love it! According to one recent survey, three-quarters of all British people start the day with toast. Nine out of ten of them eat bread every day and over half the British population describe it as their favourite food. But the British eat less bread today then any other nation in Europe. The Italians and the Belgians come top of the league, eating 80 kg per person per year!

Bread is one of the oldest foods known to man. It has been eaten around the world for over 5000 years as a staple, and its use has been linked with some of the most important events in history. The Roman conquest of Egypt was carried out to provide access to the bountiful wheat fields of the Nile Valley and Northern Africa. Ancient Egyptians paid their officials with bread and by the Middle Ages fine white loaves were considered to be food fit only for nobility in most parts of Europe. Bread has also been used as a powerful status symbol for different peoples at various times. This 'staff of life' came to be used in only its finest, whitest form for celebrating Roman Catholic Mass, and one of the world's major religious disputes centres on whether bread is to be swallowed as the body and blood of Christ or as its representation.

Whatever type of bread we eat today, whether Jewish bagel, Irish soda bread or Asian nan, none of it could be made without the fruit of different grasses; cereal grains. Various different cereals contribute to the great variety of bread types available today. One of the most

BREAD

> The bread I eat in London is a deleterious paste, mixed up with chalk, alum and bone ashes, insipid to the taste and destructive to the constitution . . .

Such was the state of London's white bread in the eighteenth century according to Tobias Smollett, the contemporary British author, who voiced this view through his fictional character Mathew Bramble.

Today's bread has come a long way since this account was written — at a time when white bread, though heavily adulterated, was reserved mostly for the rich. Nutritionists now emphasize the benefits to be gained from eating bread in most forms, but especially when made from wholemeal flour. White bread generally excludes the germ of the wheat grain (the embryo of the young plant) containing protein, oil and vitamins, the outer seed coat (the bran) which is a natural source of roughage and an intermediate layer called the aleurone, which is rich in protein, minerals and vitamins. The part ground into flour is the starchy endosperm, the food reserve of the young plant.

Today many bread flours, both brown and white have synthetic B vitamins added to them like niacin, as well as milk solids and minerals such as calcium and iron, to replace nutrients lost in the milling process.

Commercial baking is quite a different operation from home baking and involves specially chosen ingredients. Glucose syrup, hydrogenated vegetable oils, soya flour, emulsifiers (claimed rather mysteriously by some manufacturers to 'improve the eating qualities of bread') and ascorbic acid or vitamin C are common ingredients in British bread, especially that made with a proportion of soft flour. But one basic ingredient remains the same — yeast. A single-celled fungus (*Saccharomyces cerevisiae*) considered until recently to be a plant, yeast was first used by Egyptian bakers to raise bread dough and is still an essential component of most brewing and baking procedures.

Enzymes in the fungi convert sugar added to the dough mixture to carbon dioxide by a process of fermentation, and in the form of bubbles this gas gradually lifts the dough. It is the flour's gluten, a mixture of proteins which gives the dough its elasticity, allowing the bubbles to raise the bread without collapsing. Rye flour does not contain this gluten structure and so cannot be risen in the same way. For this reason many rye breads have wheat flour added to them.

prominent is rye (*Secale cereale*), which grows further north than any other cereal and on land too poor for most. This is now produced in the largest quantities in Russia, Poland and Germany, where it is milled into flour or used in its whole, 'cracked' or flaked form as the main ingredient of a number of traditional recipes. Pumpernickel from Germany (which itself boasts over 200 different kinds of bread) white and dark loaves which are often flavoured or seed-coated, sourdough and Scandinavian crisp breads are all made from a base of rye flour.

Fields of wheat in Canada

WHEAT

There are over 20 000 cultivars of bread wheat, giving great variation in shape, size and colour of wheat grains and ears. 'Hard' wheats are generally richer in protein than 'soft' and are particularly useful for bread-making. Cakes, pastries and biscuits are usually made from 'soft' wheats which contain more starch. Wheat cultivars are very versatile and are grown in almost every latitude between the Arctic and Antarctic circles. Rivet, cone or English wheat (*Triticum turgidum*) was once the principal wheat grown in southern England, but the flour milled from it is soft, so much British bread is made from imported 'hard' Canadian wheat.

The Spanish first took wheat to the North American continent in 1520. Wheat production is now so successful that some American farmers are being paid not to produce the crop, whilst in Europe a massive 'wheat mountain' has grown out of surplus grain. Yet scientists are constantly trying to produce higher-yielding varieties of wheat, with resistance to attack from diseases such as rusts, fungi with very complex life-cycles that attack the fruiting stalks and leaves. The Romans offered sacrifices of red dogs and red wine to try to appease the grain deity they held responsible. Today we turn to agricultural agencies and genetic engineers for an answer.

PASTA

Pasta, the traditional Italian staple used in a great assortment of shapes for a wide range of cooked dishes, is made from hard durum or macaroni wheat (*Triticum durum*).

The grain which supplies most of the world's bread is wheat. Eaten not only in this form, but in every sort of pasta and a huge number of farinaceous foods including cakes, biscuits, pastries, pies, sauces, coatings, toppings and fillings, this one species of grain is now the staple diet of one third of the world's population, making it in the reckoning of many scientists, the most important crop on earth.

Four hundred million tonnes of wheat are grown each year on more than 200 million hectares of land. Wheat provides 20 per cent of the world's calories and 50 per cent of its protein, and many essential nutrients as well.

There are an enormous number of different kinds of wheat. The one we are most likely to eat in our toast today is *Triticum aestivum*, or bread wheat. This is itself a vast hybrid group of wheats of which there are literally thousands of varieties.

Modern mass-produced bread wheats are quite different plants from those first gathered from the wild by our ancestors. These plants are thought to have evolved around 8000 BC from wild species of *Triticum* and the related genus *Aegilops*, in South West Asia and the eastern Mediterranean. Wild wheat ears with their long spikes or 'awns' (which help the grains to find a foot-hold in the earth) are very brittle and shatter on touch, but this and other 'undesirable' characteristics have been selectively bred out of the commercial wheat crops grown today. In the northern hemisphere Russia, the USA, China, India and France are the major producers of wheat, whilst Argentina and South Africa are the southern hemisphere's principal exporters.

A spread from plants!

The first thing that most of us spread on our bread in the morning comes either directly or indirectly from plants too. Butter — and every other dairy product, including yoghurt and cheese — is made of course from the milk produced by cows and other grazing animals, but it is flowering plants and chiefly grasses that sustain them everywhere.

Most margarines on the other hand are based on oils pressed directly from the seeds of a number of important economic plants such as the African oil palm, soya, sunflower, peanut and safflower, grown in different regions of the world.

BUTTER A quick look at a field in Britain in which dairy cattle are grazing today will tell us that the grass the cows are eating is very uniform. Thousands of hectares of what might otherwise be much more heterogeneous pastureland is now sown with one or two plant species, strains of which have been developed over a period of years to maximize the production of milk.

COW TREES FROM THE AMAZON

Some rather unlikely, but none the less possible alternatives to milk are exuded form the barks of various tropical trees, known colloquially as 'cow trees', native to South America. Early explorers to the Amazon, such as Alexander Von Humboldt and Richard Spruce noticed that the white, sweetish-tasting latex which oozed from the bark when it was cut could be drunk just like milk, without any harmful effects. The best known of these trees include *Brosimum galactodendron* and *B. utile* which belong to the breadfruit family.

Brosimum galactodendron

The principal plant involved is ryegrass. Two species are widely sown: perennial ryegrass (*Lolium perenne*) native to Britain as well as mainland Europe and parts of Asia (a naturally dominant grass in many lowland areas) and Italian ryegrass (*Lolium multiflorum*) a species introduced to Britain, which grows prolifically and has a high nutritive value. Left to its own devices, all grassland is in fact a community of widely differing species and varieties of plants; an early stage in the succession of plant colonizers of a bare patch of earth that given the right conditions would eventually succeed to shrubby vegetation and in time, forest. There are many different types of grasslands, but in the central European region, very few of these are natural. Whilst altitude and different soil types and conditions will tend to determine the flora that is present, the majority are maintained as grasslands by man or grazing animals. Almost all meadows and pastures are artificial too (in the sense that they are continually cut or grazed), but certain plant species have become adapted to these regimes and in association with one another, can be identified as distinct vegetation types.

Some of the most common of European pasture grasses include the bromes, fescues, meadow tussock and oat grasses, sweet vernal grass and Yorkshire fog. They are often accompanied by wild plants such as yarrow, daisies, buttercups, vetch, dandelion, chicory, cow parsley, hogweed and hawkbit, which together help create idyllic scenes that have come to symbolize a much-loved aspect of the countryside.

Hay made from meadows containing a large number of these plants which are collectively rich in minerals and other nutrients is always relished by grazing animals — especially sick ones which may refuse otherwise good quality hay. But of the 10 000 or so known species of grass, only a dozen have been regularly sown at different times as pasture or forage grasses.

English meadows contain a wide variety of different grasses

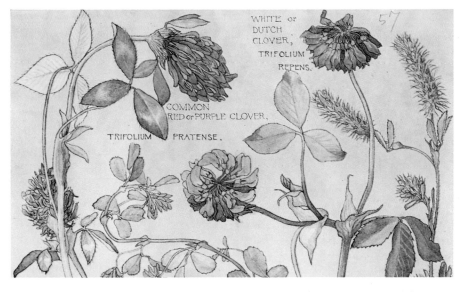

Clovers (Trifolium spp.) are often sown with grasses in pastures as they provide nitrogen compounds and improve the forage value

The most well known of these, which have helped create some of the most productive permanent grassland include cocksfoot (*Dactylis glomerata*), timothy (*Phleum pratense*) and meadow fescue (*Festuca pratensis*) as well as the now essential ryegrasses.

In today's sown pastures, especially where Italian or perennial rye grass forms the dominant cover, it is usual for farmers to include one or more clover species in their seed mix. Clovers are richer in protein over a longer period than grasses and increase the food value of the herbage. As legumes they greatly assist the plants they grow with since nitrogen compounds (necessary as fertilizers) are produced by bacteria housed in special nodules on their roots, and are released directly into the soil as the roots decay.

Until relatively recently, hay was the normal winter food for dairy cattle, but because wet weather makes it an uncertain crop, silage is now very widely used. Packed into large silos or individual plastic sacks, the action of bacteria ferments the grass and effectively pickles it, keeping it in a usable condition for many months.

MARGARINE There is hardly a processed food eaten today that does not have some edible vegetable oil in it. Around 140 million tonnes are traded around the world each year — much of it used in cakes, biscuits, pies, confectionery, salad dressings, snacks and for frying. But the largest use of edible oils in Britain is in margarine (and shortening for pastry) which accounts for some 500 000 tonnes per annum.

With vegetable oils now accounting for two-thirds of refined edible oil production in Britain, alternatives to butter are available today in a wider range of products than ever before. Relatively recent processes for the thickening of plant oils by hydrogenation (the bubbling of hydrogen gas under pressure through the oils in the presence of a catalyst such as nickel) have made it possible to incorporate many which would otherwise be unsuitable for use in margarine and other foods. Other basic procedures involving caustic soda, fullers earth or charcoal and steam treatment, for bleaching, refining and deodorizing, make quite different oils indistinguishable and therefore interchangeable, since their taste and other characteristics can be entirely removed.

The major vegetable oils used for margarine making include palm and palm kernel, soya, sunflower, rape seed, coconut, peanut, cottonseed, corn and safflower. With the exception of palm oil which is pressed from the fleshy fruit layer (the mesocarp) surrounding the palm kernel, all these oils are crushed from seeds. The use of cottonseeds has been significant as attempts to process their oil led to major advances in the industry as a whole. The ginning of cotton produces tonnes of seeds every year, but until the beginning of this century cottonseed oil was considered inedible because it contains a bitter pigment, gossypol. Experiments

PLANTS FOR MARGARINE

Sunflower (*Helianthus annuus*)

With an oil content of around 40 per cent, sunflower seeds which have a solid black outer coat, are important sources of oil for soft margarines, domestic cooking oils and salad dressings. Striped seeds are used for confectionery purposes or eating directly.

Over 3 million tonnes of sunflower seeds were grown in the EEC last year, the main producers being France and Spain. Throughout the twentieth century however, Russia has dominated world supplies, and is famous for breeding plants with giant inflorescences each of which may have up to 1000 seeds.

The sunflower is a native of North America where wild populations still occur. Here the oil from the seeds has been used to power farm machinery when mixed with diesel, and studies indicate that this mixture performs better than diesel if the sunflower oil is refined.

African oil palm (*Elaeis guineensis*)

African oil palm is native to tropical West Africa and in some areas may be found growing in large natural groves. Since the trees will thrive equally well in other areas of high rainfall within the world's humid equatorial belt, many countries now grow them in large man-made plantations — often cut from virgin jungle — and their oil has become a major revenue earner.

Large fruit bunches borne by the female trees after approximately five years contain up to 200 individual fruits. These turn from green to orange-red (or black in some varieties) when ripe. Between two and six bunches are produced by the trees each year and are harvested either by climbers using ropes, or by men on the ground who cut the fruit bunches from the crown with a sharp blade on the end of a long pole.

Whilst palm oil, which tends to be used commercially for soap making, is taken from the oil-rich mesocarp layer of the fruit, palm kernel oil which is largely used for the manufacture of edible products such as margarine, is pressed from the seed. In many areas where the palms are grown, local people press out their own oil by fermenting the fruits for a few days, then boiling and pounding them. The resulting pulp is stirred in water and the oil, which has a characteristic red-orange colour derived from the carotenes present, is skimmed off as it floats to the top.

The major producing countries of palm and palm kernel oil are Malaysia, Indonesia, Zaïre and Nigeria.

Safflower

Sunflower

African oil palm

Safflower (*Carthamus tinctorius*)

Safflower plants were originally cultivated for the deep red-yellow dye produced by their colourful thistle-like flowers, and appear to have been domesticated in the region of the Eastern Mediterranean.

The oil crushed from the seeds, however (a process mostly carried out in Mexico and India) has the highest linoleic acid content of any known seed oil. Until recently, safflower oil was used commercially for paints, varnishes and resin, but it is now exploited as a source of linoleic acid (one of the few fatty acids that cannot be synthesized by man, and important in the human diet) and for margarines and salad dressings.

involving fullers earth and steam eventually succeeded in purifying the dark coloured semi-toxic oil into a colourless, tasteless and harmless one, and techniques for hydrogenation and fractionation (fat-splitting) followed.

The various seeds processed contain varying proportions of oil — each cottonseed for example is made up of around 35 per cent oil, whilst the soyabean has only 18 per cent. Some of the highest concentrations are to be found in the coconut (65 per cent) and the fruit and seed of the African oil palm (around 50 per cent), both native to the tropics. The quantities of oil extracted from these two palms have made them important suppliers to the food industry in general, but one aspect of their chemical composition has led to the adoption of other plant oils for some culinary uses, including margarine.

Both coconut and African palm oils are highly saturated — in fact more so than butter or lard. The move towards healthier eating patterns has drawn attention to the merits or otherwise of unsaturated rather than saturated oils or fats. Since those that are saturated have been linked with cardiovascular disease, it is ironic that milk substitutes such as 'non-dairy creamers' and artificial whipped toppings, often used by those who may be worried about their fat intake, are generally made from coconut or palm and palm kernel oils. This fact and the extensive use of these oils in many other processed foods is not always readily apparent. A bill put forward in the USA requiring all food manufacturers using 'tropical oils' in their ingredients to label them 'saturated fat' has so far been defeated in the Senate.

At the other end of the scale, the oil crushed from safflower seeds has received much attention recently as it is the most unsaturated of the commonly used edible oils. It is now, along with sunflower seed oil, an important constituent of margarines and is widely used for cooking.

One famous oil not made into margarine but relished for general cooking purposes, and especially for salad dressings, is olive-oil. Like palm oil, it is the fruit of the plant which is pressed to produce this precious liquid, held in the highest esteem by peoples of its native Mediterranean region for thousands of years. Olive-oil is also highly saturated (which ensures a long shelf-life) but unlike most commercial oils today, it is rarely refined to tastelessness and remains much sought after by gourmets. 'Extra virgin' olive-oil — considered the best — is the result of the first cold pressing of the fruits, without the heat used subsequently to extract oils of lower grades. The rich greenish yellow colour of olive-oil helps distinguish it from other refined and processed oils used for margarine. The uniform pale yellow appearance of many of our margarines is due to the addition of a red-orange dye (E160(b)) processed from the seeds of *Bixa orellana*, a shrub from tropical America.

Unlike many of us who now eat the red-orange dye (annatto or E160(b)) produced by the seeds of *Bixa orellana* in our butter, cheese and margarine, the Colorado Indians of lowland Ecuador have traditionally used a thick red paste made from the waxy outer coating of the seeds to create their distinctive helmet-like hairstyle. Urucu or achiote, as the dye is also known, is widely used by many Indian groups for painting the body and for colouring threads, ceramics, implements and weapons.

Colorado Indian wearing Bixa orellana *paste on his head*

Seville orange (Citrus aurantium)

CITRUS FRUITS

The fruit of all *Citrus* species, known botanically as a hesperidium, is a berry with a leathery skin, formed from the inner pith (mesocarp) and outer layer (exocarp). This outer layer is covered with tiny pockets containing aromatic oils. Present in the leaves and other vegetative parts of the plant, they are responsible for the distinctive aroma released when peel is crushed or grated, and are widely used in cooking and perfumery. The Bergamot orange (*Citrus bergamia*) is grown almost exclusively for the oil it yields, which is much used by perfumers.

Though we expect orange peel to be a standard orange colour, in some countries where temperatures never get cool, oranges remain green even when mature. Coolness promotes the release of orange pigments, the carotenes, and if temperatures fluctuate the fruits may alternate from one colour to the other. To overcome this 'problem', some batches of oranges are treated with ethylene which promotes ripening and the development of a uniformly 'orange' appearance.

The part of the *Citrus* fruit we eat — the endocarp — is actually formed from fleshy hairs that act as juice sacks. They contain solutions of sugars (glucose, fructose and sucrose) and acids, as well as vitamin C and the vitamin B complex, carbohydrates, minerals and other nutrients. With the increasing destruction of the Asian forests which are their natural home, the wild relatives of oranges, tangerines, grapefruit, lemons and limes are now endangered.

Grapefruits (*Citrus x paradisi*) are thought to be the result of a spontaneous cross between the pummelo (*Citrus grandis*) — the largest of all citrus fruits, much eaten in its native Thailand — and the sweet orange (*C. sinensis*), an event which probably occurred only a few hundred years ago in the West Indies. The variety most commonly grown, which has a pale yellowish pulp, is 'Marsh's seedless'. Important producing areas are Florida, California, Cuba, the West Indies, Israel and South and Central America.

Citrus fruits

Citrus fruits have now become an established part of many Western breakfasts in the form of orange juice, grapefruit halves and marmalade. The production and processing of citrus fruits (for juice, peel, pulp and oil) is certainly very big business today and it is hard to imagine that only 100 years ago the orange (the 'Golden Apple' of Greek mythology) was considered an expensive luxury in North America and northern Europe.

The genus *Citrus* (Rutaceae family) comprises not just oranges but all their many relatives; lemons, limes, grapefruits, citrons, tangerines (or mandarins) and a considerable number of hybrids: the ugli for example, (a cross between a grapefruit and a tangerine), the ortanique (a hybrid orange) and the citrange (a cross between an orange and a citron).

The best known and most widely marketed today are the orange, lemon and grapefruit. The sweet orange (*Citrus sinensis*) which thrives in climates with abundant sun and reasonably dry air is thought, along with other cultivated citrus fruits, to be derived from species native to South East Asia. It is now the most widely grown citrus fruit in the world, and forming vast, regimented plantations in California and Florida is the USA's largest perennial fruit crop. Overall however, most oranges for world markets are grown in Brazil, with Italy, Spain, Mexico, Australia and South Africa also major producers.

The Arabs were the first people to mention oranges in their writings, and our word for the fruit is derived from the Sanskrit name they adopted. The Moors, who used oranges medicinally and in religous services, brought them to Spain and from there they made their way to the countries of the 'New World'. The high proportion of vitamin C contained in oranges as well as limes (also introduced to Europe by the Arabs) gave effective protection against scurvy, and this promoted their spread around the world. They were so prized in northern Europe that special greenhouses or 'orangeries' were built by the wealthy (to protect the trees from winter weather) in the hope of a fruit supply of their own.

Three main hybrid types of sweet orange are distinguished: the navel, the normal and the blood orange. The 'Valencia' is the most widely grown of the normal types — and its richly flavoured juice sets the standard by which other orange juices are judged. Though 'freshly squeezed' juice is now quite widely available in Europe, much of the packaged orange juice sold is made from filtered, freeze-dried concentrate, reconstituted with water and controlled amounts of pulp and peel. The benefits of drinking these juices are perhaps not quite as certain as those to be gained from drinking freshly-squeezed juice.

Marmalade, and other orange preserves (as well as some liqueurs) are made exclusively from bitter or Seville oranges (*Citrus aurantium*) which have a taste too sour to make them pleasant to eat raw. For European consumption many of these oranges are grown as the name suggests, in southern Spain.

Pectin, the essential setting agent needed to make many preserves and jams is itself extracted for commercial use from the peel of lemons, oranges, limes and grapefruit where it is naturally present in large quantities. The peel and pulp of apples is the other major commercial source of pectin.

Sugar

Sugar is of course essential to marmalade and jam manufacture, and it is likely to be present on the breakfast table not just in a bowl of its own, but in much of the other food and drink we consume during our first meal and throughout the day. Between us we add vast amounts of sugar to our tea and coffee and sprinkle it again on many breakfast cereals that already contain it.

All green plants make sugars and these can be stored in their fruits, roots, bulbs, stems or flowers. At different times people throughout the world have used the sap or juice extracted from plants and especially their stems to obtain sweet syrups — the date palm, sorghum and the sugar maple are notable examples. Today, however, just two plants supply most of the world's sugar: the sugar cane (*Saccharum officinarum*), a tropical grass, and the sugar beet (*Beta vulgaris*), which is related to the garden beetroot. The refined sugar (sucrose) produced from both these sources is identical and though sugar cane has supplied the world for much longer (in India people were making sweets and sherbets from it 5000 years ago), sugar beet now provides nearly half the commercial supply.

Sugar as a whole is much maligned and generally regarded as 'bad for us', but it is really only the quantity that many Westerners consume (around 40 kg per person per year in the USA and Canada) that has health experts worried. We need sugars to maintain body temperature and provide energy, but most of our essential supply is metabolized by the body itself from carbohydrates we consume such as starch, much of which is contained in quantity in cereal grains and edible roots. Sugars are also present naturally in small amounts (in the form of fructose in most green leaves and fruits and glucose, found especially in grapes and vegetables) in many of the foods we eat.

SUGAR CANE Sugar cane grows wild in New Guinea and the species cultivated on a huge scale today is thought to have been spread throughout the Pacific, South East Asia and the Middle

SUGAR

In many tropical countries, such as Ecuador, sugar cane is processed by individual growers using very simple equipment to produce a crude brown sugar. Known locally as 'panela' it is sold in block form. Mauritius, Barbados and Guyana are all important producers of raw sugar. Brown 'Demerara' sugar is named after the Demerara River in Guyana which runs through this cane-growing country.

Sugar cane (*Saccharum officinale*)

Sugar canes, which are planted from stem cuttings, may grow to 8 metres in height. Today's modern cultivars have been genetically engineered to give optimum yields under different cultivation conditions. Sugary cell sap accumulates slowly in the pithy centre of the canes and at harvesting time the pith cells will contain around 15 per cent sucrose — sugar formed as a result of photosynthesis. In many countries lengths of sugar cane stem are chewed to release their sweet juices, and are sold as a cheap alternative to shop-bought confectionery.

Bagasse, the name given to the spent cane stalks, is used for a variety of purposes, including the manufacture of paper and plastics, fuel and cattlefood, and some building materials such as chipboard.

Sweets and sherbets from India

The people of India who were growing sugar cane as a staple crop by 2800 BC made the first sweets and sherbets for their royal rulers. Lumps of processed sugar, 'Khanda' (from which the term 'candy' is derived) were either eaten alone or often mixed with rose water and ice (carried from the Himalayas) and crushed to make sherbet.

Sugar maple (*Acer saccharum*)

Before the general availability of sugar in Europe, the only practicable sweetener was honey. Sugar was first prescribed as an expensive medicine, generally to help disguise the taste of bitter herbs. In North America however, Indian groups had long used the sugar maple as a source of sweetness. After slashing the bark of the trees in the spring the sap collected was boiled down to make a thick syrup. The practice is continued on a commercial basis today and maple syrup has become a famous Canadian export and tourist attraction. The sugar content of sugar maple sap is much lower than that of sugar cane; it takes about 40 gallons of sap to make one gallon of syrup and the price remains high.

Collecting sugar maple sap in Quebec, Canada

Sugar cane being sampled in Kenya

Sugar beet

East by human migration many centuries ago. Although sugar from cane grown by the Arabs was in demand in Europe by the twelfth century as an expensive luxury, it was not until the establishment of plantations in Brazil and the West Indies early in the sixteenth century (following the introduction of the plants to the New World) that Europe had its own secure supply. Made possible only by the exploitation of generations of slaves shipped from Africa, entire countries were given over to sugar cane cultivation and huge areas of tropical forest destroyed to make way for this new cash crop. Though processing methods are now greatly updated and scientifically controlled, the cane crop itself is still very labour intensive. Much of the world's supply (around 95 million tonnes of sugar) comes today from Cuba, Hawaii, Puerto Rico, Brazil and India, where cutting of the sugar cane by hand is still common.

Despite confusion over the properties of white and brown (raw) sugars, there is nutritionally very little difference between them. Sugar is refined to various degrees of whiteness in several stages, and partly by filtering through charred bone or kieselguhr, a porous rock formed by the fossilization of microscopic yellow-green algae common both in sea water and the soil. Whilst raw sugar still contains many of the chemicals (such as protein and fat) found in the cane stalks it is claimed that none of these are of any marked benefit to the body. Molasses, the excess sticky brown liquid which is separated out from the newly formed raw sugar crystals by means of a centrifuge, is sometimes re-added to white sugar to make it brown. It can of course be bought separately as molasses or black treacle.

SUGAR BEET The Romans grew sugar beets as vegetables many centuries ago, but their use as a commercial source of sugar dates only from the beginning of the nineteenth century, following the investigations of a German chemist who had extracted 6.2 per cent of the sugar from the roots of a white variety. Today's improved varieties and modern methods of extraction have boosted the figure to around 20 per cent and large quantities of sugar beet (a biennial root crop) are now grown in Russia, France, Germany, Poland, Czechoslovakia and the USA. One of the earliest promoters of sugar beet was Napoleon Bonaparte who encouraged the development of the crop as a way of boycotting the supply of sugar from cane grown in British-dominated colonies. France is currently the world's second biggest supplier of sugar beets!

The sugar is contained in the whitish conical roots of the plants which are harvested with an average weight of about 1 kg. To extract the sugar the roots are first shredded and then heated in running water. After the removal of impurities, the clear liquid obtained is concentrated and crystallized to give a sugar indistinguishable to that made from sugar cane. Molasses is also produced from sugar beets.

The pulp that remains after the beets have been processed (along with the rosette of leaves from each plant) makes excellent food for cattle.

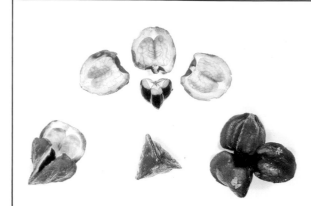

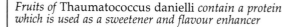

Fruits of Thaumatococcus danielli *contain a protein which is used as a sweetener and flavour enhancer*

ALTERNATIVE SWEETENERS

Sucrose, whether from sugar cane or sugar beet, has faced much competition in recent years because of the introduction of liquid sweeteners made from other sources.

Chief among the rivals are corn syrup which contains glucose (made from hydrolysed maize starch) and fructose, and invert sugar syrups made by treating this same starch with enzymes or acids that convert it into glucose molecules. These syrups are widely used in place of sugar in many processed foods, especially soft drinks, and alcoholic drinks and baked goods, where quick fermentation is required.

Some interesting alternatives to sweet syrups and sugar exist as natural components of other plants. Potentially much safer than the very low calorie synthetic sweeteners already developed (such as saccharine which is made from petroleum, cyclamates whose use is now largely banned, and aspartame formed by the unnatural bonding of two amino acids — the source of which is kept a closely guarded secret), three plants in particular, all from tropical West Africa, have aroused much scientific interest.

The first of these, a herbaceous plant that grows up to 3 metres high, *Thaumatococcus danielli*, has produced what is probably the sweetest substance known to man — up to 4000 times sweeter than sucrose! Crimson fruits which develop just above the soil surface contain the substance in the soft jelly-like aril that surrounds the black seed.

In the humid rainforests of its native West Africa, *Thaumatococcus danielli* is well known to local people. They use its broad, flexible leaves as disposable plates or for wrapping food prior to cooking, and the long thin leaf stalks are harvested for mat-making. Children suck the sweet arils of the fruit and these are also used as a source of sweetness in food preparation. The protein (thaumatin), now extracted for commercial use from the fruits is effectively non-calorific, and therefore very suitable for dieters and diabetics. In Japan it appears widely in chewing gum, soft drinks, canned coffee, flavoured milks and even cigarettes. It is also sold as a flavour intensifier, since it will enhance a variety of sweet and savoury tastes (including peppermint and coffee) while suppressing any bitterness.

The bright red fruits of the serendipity berry (*Dioscoreophyllum cumminsii*) which are only about one centimetre long and grow in grape-like clusters of 50 to 100 fruits, are also intensely sweet. The active principle, a protein called nonelin, is around 3000 times sweeter than sucrose. The fruits of a third West African plant, *Synsepalum dulcificum*, have the unusual power — in the words of the Kew Bulletin, which published details as long ago as 1906 — 'to change the flavour of the most acid substance into a delicious sweetness'. The taste of a lemon or some other sour substance becomes wonderfully sweet after chewing just one of the plum-like fruits for a short time. The glycoprotein responsible (aptly named miraculin) which has been extracted from the fruits, has undergone much investigation for use as a food supplement in the West.

The Japanese are the pioneers of another plant-derived sweetener, 'stevioside' — a white crystalline powder 250 to 300 times as sweet as sucrose — refined from the leaves of *Stevia rebaudiana*. The plant, a member of the Compositae family, has long been used for sweetening drinks by the Guaraní Indians of Paraguay, who call it Caa-ehe.

Lippia dulcis from Mexico, which was known to the Aztecs, has also been the focus of intense research, though much more recently. Scientists at the University of Chicago have isolated from the leaves and flowers and further synthesized a compound which they say is 1000 times sweeter than sucrose.

A bowl of cereal

To most of us, a bowl of cereal means cornflakes, muesli or some other mixture that arrives on our breakfast tables in attractive boxes, needing only the addition of milk, and perhaps sugar. The main ingredients are, as the names tell us, derived from cereal grains such as maize, wheat, rice and oats, often in the company of nuts and soft dried fruits. However, these grains are often (especially in non-muesli mixtures) so highly processed that is is difficult to remember what it is we are really eating! To many rural peoples — for whom such packaged luxuries are unheard of — a bowl of cereal would be more likely to mean; in South East Asia a portion of plain boiled rice; in South and Central America a dish of cooked maize kernels; and in Africa a kind of porridge made from pounded sorghum or millet. Along with wheat, these cereal grains are responsible for at least half of all the calories we eat.

THE WORLD'S CEREAL BOWL

Ceres, the Greek goddess of grain has given her name to the grains that provide at least half of the world's calories today. They are grown on nearly three-quarters of all the farmland on earth, feeding not only people but most of our domesticated animals too.

Cereal grains are generally considered to have been a prerequisite for the development of the world's major civilizations: maize for those of South and Central America; wheat and barley for the Near East and Mediterranean Basin and rice and millet for the populations of the Far East.

Terraced paddy field in Kashmir, India

RICE

With some 1.6 billion people dependent on it as their staple food, rice could be described as the most important crop in the world. Two hundred million tonnes of rice are eaten each year, mostly in Asian countries. There are at least 20 different species of *Oryza*, but only *O. sativa* (Asian rice) and, to a much lesser degree *O. glaberrima* (African rice), are cultivated.

Because of its long history of cultivation in many countries, there are thousands of local rice varieties, but two major groups are recognized: the japonica or sativa types which come originally from Japan and Korea and the indica types from India, China and Indonesia. Grains are also divided into 'long' and 'short', and classified according to the texture of their endosperm. Plant breeding programmes, such as those underway at the International Rice Research Institute in the Philippines, are constantly improving rice varieties and developing new cultivars with very high yields.

Wild rice belongs to another genus *Zizania* native to the New World. Grains of *Z. aquatica* were an important source of food for North American Indians and are grown commercially but on a limited scale in parts of North America.

Like its wild relatives, rice is a swamp plant. Its specialized stem anatomy, which allows oxygen to reach the roots, has enabled it to be grown in paddy fields which are flooded for much of the growing season. Some varieties however, classed as upland rice, do not need to be grown standing in water as long as an abundant supply is at hand. This has extended rice growing to areas such as Australia and also Brazil — which now produces most of the world's upland rice on huge plantations cut from the jungle.

Rice was first grown thousands of years ago in India and China and is an integral part of most Eastern cultures. Many Asian peoples do not consider their meal complete without it. Our own adopted custom of throwing rice at weddings reflects the oriental view of rice as sacred and a symbol of fertility.

Most rice-based breakfast cereals use polished white grains from which the nutritious outer layers (pericarp and aleurone, as well as most of the embryo) have been removed. As a source of almost unadulterated carbohydrate, these often 'puffed' grains of rice are synthetically enriched with vitamins — especially of the B complex — lost in the milling process.

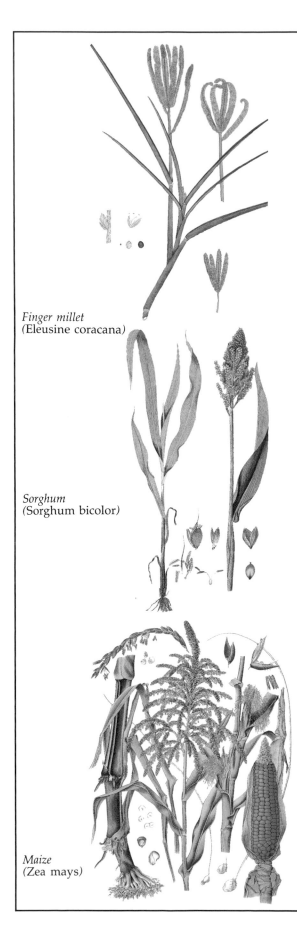

Finger millet
(Eleusine coracana)

Sorghum
(Sorghum bicolor)

Maize
(Zea mays)

MILLET

More drought-resistant than any other major cereal crop, pearl millet *Pennisetum glaucum* (*P. americanum*) has become the staple food for millions of people in India and Africa, who live on desert fringes or on very arid lands. It will give good yields in some of the driest and most inhospitable regions on exhausted and nutrient-poor soil. The cereal grains are generally ground into flour for making into flat breads, or eaten whole.

Another of the collection of grasses known as millet, *Eleusine coracana* (more commonly finger millet) is a staple of Eastern and Central Africa, though it is believed to have been domesticated in the north-eastern part of the continent. It can be stored without deterioration for up to ten years, and is largely resistant to weevil attack. Finger millet is usually eaten as porridge or ground into flour for making into large flat breads. India and China are the chief producers of millets for international trade.

SORGHUM

Another grain native to Africa is sorghum (*Sorghum bicolor*). Subject to intensive selection by native peoples over thousands of years leading to many distinctive varieties, the cultivated sorghums are generally grouped into four main types based primarily on use. These are grain sorghums, sweet sorghums (or sorgo), sudan grass (a different but related species) and broomcorn or broom millet.

Though varieties are also grown in the southern United States (chiefly for animal feed) sorghum has been a traditional grain of hot regions which receive too little rainfall for most other crops to grow. It is the major food of many African and Asian people who often grind the grains into flour for flat breads or use them to make beer.

MAIZE

Millions of British people start the day with a bowl of maize, not in the form of whole grains, but as cornflakes, the nation's most popular breakfast cereal. Stripped of the fibre and nutrients present in the outer layers and embryo of the grain, the pre-cooked starchy endosperm is rolled and toasted with the addition of a range of synthetic vitamins, malt flavouring, salt and sugar.

A great number of different myths, still told by native peoples of South and Central America for whom maize is a staple food, give to our ears naïve and implausible accounts of the origin of maize: in some a fox or little bird steals the grains from heaven, in others the moon in the form of a man brings them to earth as a precious gift. Our scientists appear to be having a much harder time explaining the origin of maize! Two giant New World grasses, teosinte and *Tripsacum*, and a recent discovery *Zea diploperennis* are believed to be closely involved, but though Mexico has been agreed on as the place from which domesticated maize spread north and south over 7000 years ago, its ancestry is still unclear.

The scientific name for maize (or corn as it is generally known in America) *Zea mays* was established by Linnaeus: 'Zea' meaning 'cause of life' and 'mays' a rendering of the name 'Mahiz', 'our mother' given to the plants by the Amerindian peoples of Cuba and Haiti.

For many people of the tropics, maize is a major part of their diet. It has about the same number of calories as wheat or rice, but though lower in protein and deficient in the amino acid lysine it is higher in fat and richer in thiamine. It is often grown in the company of beans which supply nitrogenous compounds that the plants can use.

By the time Europeans first reached the Americas, some 300 major maize varieties were being grown from Canada to Chile. Although there are now many thousands of cultivars, many of them hybrids, six main varieties are recognized today: pod, dent, flint, pop, flour and sweetcorn. Each has a distinctive use and appearance, determined by the structure of the grains. Sweetcorn, the variety eaten in the USA and Europe as a vegetable, is so named because part of the sugars in the endosperm are not converted into starch until sometime after picking, and so are 'sweet' to taste. Popcorn, developed as a crop by ancient Peruvians, contains mostly dense endosperm inside a tough coat. Dent corn which dries with a small depression or dent in the surface of each grain, and flint corn varieties were the traditional crops of North American Indians, who showed European settlers how to grow the plants successfully.

Central and South American peoples developed a number of delicious and nutritious ways of eating maize. A sweet porridge was widely consumed and tamales, a 'dough' made from the cooked, mashed endosperm, filled with sweet or savoury mixtures of beans, peppers, and meat are eaten in many parts of Latin America today. Chicha, a fermented

Andean woman with bowl of boiled maize

maize beer and tortillas, flat pancakes made from maize are other traditional uses. Corn oil for cooking or for salad dressings was a Western introduction — a by-product of the maize-milling industry.

Nearly half the world's maize is grown in North America today, and it is the country's most important crop. About 90 per cent, however, is fed to farm animals. Russia, China and South America grow much of the rest of the world's supply, but since the majority is consumed locally, only 10 per cent of the total enters world trade. Maize flour and starch have come to play a major if unsuspected role in many of the items and industrial processes we take for granted. The Aztecs used cornflour to thicken their famous chocolate drink. Today this and maize starch thicken a huge range of processed foods from sauces and instant soups to liquorice all-sorts. Maize starch is also used to make adhesives, pharmaceuticals, cosmetics, cloth and paper size, and the glucose syrups with which a very large number of processed foods and drinks are now sweetened. A vast array of other products, from paints and plastics to synthetic rubber can all now be made at least in part from processed maize.

Various other cereal grains however, that are little known or appreciated in the West, make highly nutritious foods. T'ef (*Eragrostis tef*), the most important crop in highland Ethopia, (with grains so small that seven will fit onto a pin head) and Adlay or Job's Tears (*Coix lacryma-jobi*) from India are two of these.

Several plants that do not belong to the grass family, but which produce cereal-like grains are also valuable sources of protein. The amaranths (about fifty species belonging to the genus *Amaranthus*) are an example of food plants whose virtues have only recently been 'discovered' by the West. Native to South and Central America, the Aztec ruler Montezuma received tributes of thousands of pounds of Amaranth seeds every year, and Andean peoples also held the crop in high esteem both for dietary and medicinal uses. Up to half a million tiny round seeds may be contained on one deep red seed head, weighing up to 2.7 kg. The crop is resistant to drought and can thrive on poor soils and arid land, yet its protein content is comparable to that of wheat.

Quinoa (*Chenopodium quinoa*) native to Peru, was once the cherished crop of the Incas. It was banished by the Spanish, however, when they conquered the country and is now in decline generally throughout the Andes. Its white, yellow, pink or black seeds, which contain about 15 per cent protein, are ground into flour for bread and cakes, or eaten whole in soups. They are also used to make chicha beer. The leaves of *Chenopodium* (from the Greek for 'goose foot' which refers to their shape), can also be eaten as a spinach-like vegetable.

Buckwheat (*Fagopyrum sagittatum*) is native to Central Asia, but is now widely grown in Canada and the USSR. Its three-cornered seeds which resemble beech mast, led to its German name Buchweizen (beech wheat) which later corrupted to buckwheat. In Russia a nutritious porridge (Kasha) is made from the flour.

BREAKFAST DRINKS FROM PLANTS

The Japanese bathe in coffee grounds, believing them to have health-giving properties

Coffee

Second only to petroleum as a revenue earner, coffee is an immensely valuable commodity. There is hardly a country in the world that does not drink it and around 25 million people rely on its production for their livelihoods. Without coffee, many breakfasts and dinner parties would be incomplete and social gatherings the world over would lack an important symbol of hospitality. The Japanese, who bathe in coffee grounds believing them to have health-giving properties and the Turks who scan the dregs of their coffee cups for omens of the future, would also have to look elsewhere!

Coffee berries develop in 7 to 9 months, turning from dark green to yellow then red

The drink that revives hundreds of millions of us (about one-third of the world's population) and especially the Finns who drink on average five cups each a day, is made from the roasted seeds of a tropical evergreen shrub, now grown in some fifty different countries. Though only two are cultivated on any scale, three different species of coffee bush are recognized commercially giving coffees known as Arabica, Robusta and Liberica. *Coffea arabica* (which gives Arabica coffee) is the most widely grown, producing about 75 per cent of the world's coffee (chiefly from South and Central America, India and its native Ethiopia), whilst *C. canephora* (which gives Robusta) yields some 24 per cent.

All cultivated coffees are small trees with glossy, evergreen leaves and white, sweet-smelling flowers. After fertilization, mature berries (or cherries as they are also known) develop from the flowers in about 7 to 9 months, turning from dark green to yellow then red. Inside the sweet pulpy outer layers are two coffee seeds (or 'beans') surrounded by a delicate silvery seed coat. Because coffee bushes cannot tolerate frost they are grown in tropical and sub-tropical countries with an average rainfall of at least 1.9 metres per annum.

The discovery of coffee is attributed in legend to an Ethiopian goatherd who noticed that his goats had become unusually frisky after eating the ripe red berries of wild coffee bushes that grew in the forested mountains of the country. The goats' activity was due to their ingestion of caffeine, a natural stimulant contained in the coffee plant. By the second century AD, local tribesmen were making small cakes from the pulverized fruit mixed with grain and animal fat to sustain them on long journeys and to relieve fatigue. The berries were also fermented and mixed with water to make a stimulating drink.

The Arabs, however, were the first people to brew coffee as we know it today, and by the thirteenth century, with the opening of coffee houses (where people from all walks of life could gather and socialize) the drink had become extremely popular. Coffee drinking spread quickly throughout Arabia and into Turkey and the surrounding areas. To offer any visitor a cup of the strong black brew became an important gesture of hospitality.

The Arabian drink did not reach Europe until 1616, but it did not take long for coffee houses to become all the rage. By 1675 there were over 3000 in London alone. Dubbed 'penny universities' since it was possible to exchange gossip and learning there for the price of a cup of coffee, and as forums for political debate, they were regarded by anxious politicians as centres of sedition. Attempts were made to close the houses down, but such an uproar ensued that the idea had to be abandoned. The English Stock Exchange, the merchant banks and the first insurance companies all had their beginnings in coffee houses.

Coffea arabica

COFFEE

The continuing destruction of Ethiopia's montane forests — about four-fifths have already gone — now threatens the survival of the wild coffees from which *Coffea arabica* developed. The genes of these wild trees have already proved invaluable in helping our cultivated coffees fight the dreaded coffee rust disease. In 1970, Latin American plantations were saved from devastation, and national economies from disaster — with genetic information from a rust-resistant strain found in the forests of Ethiopia. In patches of forest on Madagascar, wild coffee varieties have been found that are completely free of caffeine.

Inspiration in a cup

Unlike Beethoven who, it is said, measured out exactly sixty coffee beans for every cup of coffee he brewed, most British and Australians prefer a quicker, simpler drink — painstakingly prepared by someone else!

To make a jar of instant coffee, coarsely ground beans are put into huge, sealed stainless steel percolaters and brewed under pressure for several hours. Since the liquid is constantly recycled at very high temperatures, coffee aromas are added to make up for the loss of the subtler flavours.

This concentrate is then sprayed under pressure through fine nozzles into a very high tower. As it falls, the liquid dries into a powder which is often tumbled with steam to induce the formation of granules.

Another method is to freeze dry the coffee brew into thin sheets which are then cut into granules. The temperature of these sheets is allowed to rise whilst under vacuum, resulting in the water present 'boiling off' at a very low temperature, without the coffee solids becoming wet. This leaves small dry particles with no heat damage and minimum loss of volatile aromatics. Unless treated by special processes involving solvents, steam or water, all the coffee currently marketed, whether freshly ground or instant, contains the drug caffeine. Caffeine is a general cellular stimulant. It increases the metabolic rate, stimulates the heart and mimicks the feelings produced when the body releases adrenalin, generally making us feel more alert. Though the inspirational effects of pure coffee are not to be denied (Bach, for example, wrote his Coffee Cantata in praise of the drink), an excess of caffeine can cause anxiety, dizziness, heart palpitations and even mild delirium!

Turkish coffee

Thick black Turkish coffee, associated with ritual and Eastern mystique, is brewed over a flame in a traditional ibrik.

Water and sugar are added to the ground beans and the mixture is brought to the boil. It is then removed from the heat and the process repeated several times until the right consistency is reached. The coffee is served very hot and frothy, in tiny cups, which soon accumulate a silty layer of grounds at the bottom.

Messages from the past and omens for the future are discerned by those able to 'read' the dregs. In earlier times, a Turkish woman could divorce her husband if he failed to supply her with coffee for every day of her married life!

Alternatives from other plants

Many substitutes for coffee have been drunk over the years. The best known is chicory (*Cichorium intybus*) which became popular during the eighteenth century, and is still favoured by the French. The bitter, fleshy taproots — brown externally, but yellowish white inside — are ground and then roasted for use either alone or as a coffee additive.

A more luxurious substitute is the fig (*Ficus carica*) which is also roasted and ground. Austrian and Bavarian coffees are well known for this addition.

Cheaper, coarser alternatives are made from rye, or barley and sometimes from peas, oats, dates, maize and acorns. Dandelion roots (*Taraxacum officinale*) are also made into a coffee-like drink — since they possess tonic and stimulant properties, but lack the caffeine of coffee beans.

Chicory (Cichorium intybus)

The great demand for coffee compelled European businessmen to look for their own supplies — enabling them to be independent of the Arabs who had monopolized the trade since it began. The Dutch were the first to secure live coffee seeds from Mocha, the centre of the Arabian coffee trade, and by the mid-seventeenth century they had established plantations in Sri Lanka and the East Indies. From these plantations bushes were sent to Amsterdam's Botanic Garden, but only one plant was to survive the difficult journey. Seeds from this plant were then sent to other European botanic gardens. Astonishingly, a single coffee bush grown from the specimens held at the Jardin de Plantes in Paris, carried by the Frenchman Captain de Clieu to Martinique in 1723, began the entire Caribbean coffee industry which later spread to coastal South America and Brazil — now the biggest producer of coffee in the world.

In Britain 90 per cent of all the coffee drunk is 'instant', made from soluble granules of a spray or freeze-dried coffee brew. Real coffee fans however, and most Americans (for whom these statistics are almost reversed) stick to the drink brewed directly from the freshly ground and roasted beans.

Much of the ground coffee available pre-packaged is made not from one single type or batch of beans, but from blends of perhaps 7 or 8 different coffees that are carefully selected to give a desired taste. The variety of coffee bean, the region it is grown plus the method of its preparation and subsequent roastings will all combine to produce a distinctive appearance and aroma.

After picking, mostly by hand, the sweet pulp which surrounds the coffee beans (really seeds) which develop inside each cherry-like fruit is removed in one of two ways. The dry process, which produces beans sometimes known as 'hard', 'native' or 'natural' in the trade, involves first drying the whole fruits in the sun or artificially and then removing the dried pulp, the fine protective endocarp or 'parchment' and the silvery skin surrounding the seeds. In the wet process, which produces 'mild' coffees, with a superior flavour, the fruits which are usually only picked when ripe are first de-pulped by machine, exposing the thin parchment then washed and left to ferment for 12 to 24 hours. During this process (which will loosen the parchment) a chemical alteration takes place in the beans producing substances that will eventually develop into the characteristic coffee aroma and taste. The beans will then be dried in the sun for about a week, whilst still encased in the parchment and their delicate semi-transparent silver skins. These layers will not be removed until just before shipping.

Roasting, which reduces the moisture in the dried green beans and brings out their aromatic oils, is the all-important final stage, developing the optimum flavour and appearance of the beans and determining the smell of the coffee. Very sophisticated computer-controlled machinery now roasts most of our beans with hot air at the touch of a button. Great experience is required by those controlling the machines, to reach exactly the right roast — a few seconds either way will make a great deal of difference!

In the trade, coffee which is unblended, that is from one particular area or plantation, is referred to as 'pure' or 'original'. There are several hundred of these coffees on the market today — a gourmet's choice for everyone!

Coffee beans being dried in the sun, and sifted

Tea

Would you say your cup of tea was autumnal, pungent or brisk? Hopefully it will not be weedy or chesty!

These are just some of the terms used regularly by Britain's professional tea tasters as they sample teas shipped to London from tea plantations all around the world, for sale by auction. It is the tea tasters who will decide which of more than 3000 different types or grades of tea will enter our homes in tea bags and which in more expensive packets and tins.

BUYING AND SELLING TEA

Tea auctions are held in the City of London every week — the only place in the world where teas from every tea-producing country will be on sale. It is here that the large retailers of tea bid for the different teas available, buying in lots of twenty to one hundred chests at a time. Tea-tasting is a vital part of the buying and selling process. Special pots are used to brew the various sorts. First the brokers, who act as agents for the large tea estates try the teas and suggest a price. The tasters for the retail trade then also taste the teas and decide which ones to buy and what they will pay for them.

The auction room at the East India Company

Over two million tonnes of tea are produced each year from approximately 25 countries. Asia is the largest producer, providing 79 per cent of world production in 1986. India is the largest individual tea-producing country, with nearly 30 per cent of the world's tea grown there. Much of this is now consumed within the country however, and although Sri Lanka grows only 10 per cent of the world's tea it exports about the same amount as India. East Africa, which began commercial planting in the 1920s is also now a major producer, with Kenya exporting some very fine quality teas. Indonesia, Malawi, Thailand, Mozambique, China, Turkey, Tanzania and the USSR are also important producing countries.

Last year North Americans drank about 35 billion cups of tea made from some 68 million kg of tea leaves, but the British are the world's biggest tea drinkers, consuming 20 per cent of the world's tea imports each year. The drink that is now enjoyed by over half the world's population is made from the leaves of a small evergreen shrub related to the garden camellia, *Camellia sinensis*.

EARLY HISTORY AND USES *Camellia sinensis* is thought to have originated in Tibet or Central Asia, but the plant is indigenous to a large fan-shaped area bordered to the north-west by Assam, to the north-east by the Chinese coast and to the south by southern Cambodia and Vietnam.

The earliest uses of tea were almost certainly medicinal. The Chinese, who have been cultivating tea for more than 2000 years, used it as a cure for abscesses and tumours, ailments of the bladder and inflammations of the chest, though they also noticed that it quenched the thirst and kept them awake!

The Chinese believe that tea was first drunk during the reign of a legendary emperor Shen Nung (The Divine Healer), who lived about 2737 BC, but the present word for tea, Ch'a, was not used until the seventh century AD. Before this the names of several other shrubs were used interchangeably.

JAPANESE TEA CEREMONY

By the tenth century, tea was being taken in the form of a thick green froth made by whipping dried and finely powdered tea leaves in hot water with a bamboo whisk. It was at this time, during the Sung dynasty (960-1120), that the Ch'an sect, a branch of Chinese Buddhism, ritualized the preparation and drinking of tea as a means of contemplation.

They took this ritual with them when they moved to Japan in the thirteenth century, thus beginning the Japanese tradition of the Ch'a-No-Yu, as it was called, the Ceremony of the Hot Water Tea. Zen Buddhism as the religion now became, embraced the tea ceremony as a means of reaching the ideal state of perfect harmony and tranquillity at the heart of Zen ideology.

Today the tea ceremony is still a very important part of Japanese life, exemplifying the correct mode of conduct for the upper classes. Young Japanese girls can attend courses to teach them the skills required to lead a tea ceremony, and families may spend large amounts of money on the construction of a special tea hut and the acquisition of special tea utensils.

The first tea huts were crude, simple straw-thatched buildings entered on the hands and knees. Special gardens which aimed to represent a form of purified nature, were laid out around them. The simplicity and austerity of the huts and the formalized gardens were designed to help induce the state of harmony necessary for contemplation. The ritual arranging of flowers, carried out by the tea masters as part of the ceremony and which developed into the present day art of Ikibana (flower arranging), is also intended to portray the ideals of serenity, grace and contemplative beauty.

By the fourth century, however, there are reliable records describing the cultivation, processing and brewing of tea. The tea-cake method was then the standard procedure. A cake would be made from young leaves mixed with rice and then baked. This cake was then crumbled into small pieces and boiled a bit at a time, with slices of onion, orange and ginger to make a stimulating drink.

TEA DRINKING IN EUROPE Europe first heard of tea in the mid-sixteenth century through the tales brought back by travellers to the East, but it was not until the mid-seventeenth century that commercial tea drinking reached eastern Europe and its use did not become general until the eighteenth century.

The Dutch and Portuguese were early promoters of tea, though the French regarded it as a 'dubious drug'! British nobility, however, favoured the drink and Charles II gave the monopoly of tea importation to the British East India Company, thus freezing out the Dutch and establishing a regular run between Canton, Amoy and England.

By 1700 Britain was importing a million pounds (nearly half a million kg) of tea a year and there were several hundred shops in London serving tea to drink. Coffee drinking had also become well established by this time, but like the coffee houses, tea houses did not allow women inside. Thomas Twining came to the rescue by building an annexe onto his coffee house just for women, and in time the British pottery companies Spode and Wedgewood developed handles for the simple Chinese-style cups in use at the time so that the ladies would not burn their fingers. The British teacup was born!

By 1780, as imports of tea into Britain reached 14 million pounds (6.3 million kg) per year, coffee houses gave way to tea houses in popularity and tea became the approved drink of those who fiercely opposed the drinking of alcohol. The term 'tea-total' stems from this time.

A harsh tax was placed on tea, however, by King George III and in 1773 a serious upset occurred for the British East India Company. By way of protest a group of American colonists, disguised as Indians, threw a cargo of tea waiting to be unloaded into the harbour at Boston, an event which has been remembered as 'The Boston Tea Party'.

This heavy duty placed on tea also affected British tea drinkers and meant that smuggling became commonplace. The secret trade

Dr Samuel Johnson was a keen advocate of tea

involved a wide range of people — including wealthy merchants and respectable members of the clergy, who sometimes stowed their tea in the crypts of parish churches.

During the early nineteenth century tea was still very much a drink for the rich, and they objected very strongly to the desire of poor people to drink it too. A very unpleasant substitute was sold to the unwary made from ash leaves and sheep's dung!

By Victorian times, however, the taking of 'afternoon tea' had become a British institution. It is said to have originated in 1750 with Anna, wife of the seventh Duke of Bedford, who complained of having a 'sinking feeling' in the late afternoon.

TEA GROWING Tea is grown mainly in the subtropics and mountainous regions of the tropics. Depending on the climate, it can be grown between sea-level and just over 2100 metres. Near the Equator it is often cultivated at heights of around 1200 to 1800 metres. In Kenya and the Darjeeling district of northern India elevations of some 2100 metres produce excellent teas.

Tea plants require even temperatures with moderate to high rainfall (at least 1.3 metres per annum) and high humidity throughout the greater part of the year. They will not stand frost.

Young tea bushes in nursery beds

Woman picking tea

Young tea bushes are raised on plantations from the seeds or rooted cuttings of selected individuals and are planted in long, straight lines usually as a single crop. They are allowed to grow until they become large shrubs and are then severely pruned to encourage the formation of twigs. They are kept at about knee-height by regular pruning. Every week to ten days the tea bushes produce a 'flush' of tip growth — two very tender young leaves and an unopened leaf bud. These 'flushes' as they are known are picked by hand, mainly by local women, and thrown into baskets carried on their backs. Experienced pickers can gather up to 30 or 35 kg of leaves and buds each day.

Tea supply is very dependent of the vagaries of the weather. In some parts of the world, such as Darjeeling and Assam in India, tea-picking is seasonal, (taking place between March and October), in others — such as Kenya — it is picked all year round.

DIFFERENT TYPES OF TEA There are two main varieties of *Camellia sinensis*. *Camellia sinensis* var. *sinensis*, known as China tea, made the first cup of tea to be drunk by Europeans, and was the first tea plant to be encountered by travellers to China in the late eighteenth century. For some two hundred years teas from China had been the only ones known to the Western world, and since the Chinese had long resisted attempts by the British and other traders (whom they regarded as barbarians), to gain any established rights and trade connections there, the British in particular were keen to grow their own supply in India.

Several expeditions to China were launched to try to bring back seeds or plants. In 1793 Lord Macartney, head of the British mission to Peking, collected 'several tea plants in a growing state with large balls of earth adhering to them' with the idea of planting them in Bengal, but these unfortunately died. Other travellers had better luck, collecting seeds which later thrived in the Botanic Garden of Calcutta and which led to the first experimental plantings of tea from China around 1818.

In 1823, however, an important discovery was made. *Camellia sinensis* var. *assamica* — Assam tea, was found growing wild in the hills of Assam in northern India by two employees of the East India Company: Charles Alexander and Robert Bruce. At first, botanists thought they had discovered a new species of tea and the plant was called *Thea assamica*. General knowledge about tea was still so limited it was not known, for example, whether green and black teas came from the same plant — so tea cultivation in India did not begin immediately Assam teas were found.

The first significant shipment of Assam tea (eight chests), did not reach England until 1838, and at that time the British still thought that superior tea could only be got from China. Thus it was that in 1848, Robert Fortune, appointed collector by the Royal Horticultural Society, went out to China and successfully brought back a large quantity of seeds and 2000 seedlings for the East India Company's plantations which were now becoming established in the Himalayan foothills. But these Chinese plants did not flourish, and at last it was realized that the Assam tea, which was indigenous to the area, was much better suited to local conditions. Assam tea was subsequently cultivated on a very large scale in commercial plantations still in use today in north-east and southern India and in Sri Lanka.

At the beginning of the twentieth century Assam tea was introduced to East African countries, principally Kenya, Tanzania and Uganda, where it has now become a very important crop. Assam tea is also grown commercially in Papua New Guinea.

China tea, which was introduced to Sri Lanka for commercial cultivation around 1828, is now produced on a commercial scale mostly in China itself though Sri Lanka is still an exporter. Japan, Formosa, Mozambique, Turkey, Russia, Malawi and Iran also grow China tea.

The popular brands of tea available in the shops are a blend of many different teas, combined by experts for strength, flavour and colour to produce the same 'blend' continuously. In general, China tea leaves produce a tea with a flowery flavour that is lighter and finer than that made from Assam leaves, which make a 'heavy', stronger cup of tea. The major distinction that is made between the teas we buy, however, refers to their processing method, giving us either 'black' or 'green' teas.

Black tea makes up the bulk of the commercial tea consumed throughout the world and is by far the most common type of tea drunk in Europe and America. To make black tea, the picked leaves are first 'withered' by placing in special withering 'troughs' where air is blown through

Tea leaves being rolled to release enzymes for fermentation

Graded tea is packed into chests for shipment to auctions overseas

them. This reduces the water content of the leaves and begins the enzymatic processes that will release the tea's aroma. The leaves then enter the 'rolling' stage, passing through special machinery which ruptures the leaf cells and releases enzymes from the juice that will cause chemical oxidization of the phenolic compounds present. This process, known as fermentation, takes place in a warm, damp atmosphere until the leaves, having absorbed oxygen, become a bright, shiny copper colour — like a new penny.

The last stage is to dry or fire the tea with hot air to stop the fermentation and remove excess water. It is during the firing that the tea leaves acquire the characteristic dark colour of black tea. The best black teas include Darjeeling, Nuwara Eliya, Kenya and Keemun.

Well known grades of black tea include the pekoes, such as orange pekoe (a reference to the orange or golden-coloured leaf tips) which gives a light, pale tea, and is grown in Sri Lanka, Japan, China and Java, and the Souchongs, such as Lapsang Souchong, which originally acquired its distinctive flavour from the smoke from burning rope which was placed beneath trays of leaves to quicken the drying process. Darjeeling gives some excellent grades with charmingly coded names: TFGOP1, for example, stands for Tippy Flowery Golden Orange Pekoe N° 1!

The major difference in the production of green and black tea is that green tea does not undergo the fermentation process. The freshly picked leaves are steamed and dried without the withering stage and thus retain a faint but distinctive 'grassy' taste, and of course their green colour. Examples of green teas include Imperial grown in China, Gunpowder from China and Taiwan and Yamashiro from Japan.

Oolong tea is half-way between black and green tea. It is a very fine tea, produced by a short withering process, and is lightly fermented to give a delicate flavour of peaches.

The 3000 or so different types of tea that are distinguished by tea tasters and blenders are the result of a combination of factors, all important to the appearance and flavour of the finished product.

These include:

i) the type of bush.

ii) the processing method that the tea variety or hybrid undergoes (ie. to produce black or green tea).

iii) the size, age and part of the leaf picked. There are five recognized leaf sizes. Older, coarser leaves, for example, will fetch lower prices and produce coarser tea with less fragrant flavours and odour.

iv) the type of rolling used to begin processing. There are two types: orthodox, whereby the leaves are rubbed between large ribbed metal rollers, and CTC, which stands for cut, tear, crush, a process that breaks down the leaves into small pieces and produces the tea that fills our tea-bags.

v) the region the tea is grown in, including the height above sea level. The best teas are acknowledged to grow at higher altitudes, in the cool, mountainous areas of India, Sri Lanka, Kenya, Formosa and Japan at elevations of between 1200 and 2100 metres above sea level.

vi) climatic and soil conditions. The best soils are deep, permeable, well-drained acid soils, and are often tropical red earths.

Different types of tea can be produced by flavouring with various fruits

Bearing all these factors in mind, different plantations will produce quite different teas. The Indian term 'jat' is used to distinguish these different teas on the basis of the plantation or district that they have come from or on the basis of leaf characteristics.

The names of the teas that we can buy in the shops may refer to their place or origin, a particular processing technique, the size or condition of the leaves or even a person with whom they are connected. Earl Grey is perhaps the most famous example, remembered for his addition of oil of Bergamot to his tea.

COMMERCIAL TEA DRINKING The famous British 'cuppa' and the tea enjoyed by most Americans is generally black tea, whilst green tea remains more popular in the Far East.

Most commercial teas are blends of different types and grades of tea. There are often around twenty different teas in one blend. North Indian teas, for example (especially from Assam) are used to give strength, South Indian, Sri Lankan and Indonesian teas are used for flavour, whilst African teas give colour.

'Good' teas will tend to have a bright appearance whilst the cheaper the tea the muddier the colour. The range in prices is enormous — varying from about 50p per kg for the lowest quality tea to £20 per kg for the best. The soft, hairy tips of the tea leaves are said to produce the best and most expensive teas, but the time of picking is crucial. The second flush (that is, the second growth of leaves) produces Darjeeling with the finest flavour, whilst the first flush will produce a weaker, more delicate cup of tea.

The stimulating effect of tea is due to the caffeine it contains, whilst the flavour is a product of the fermentation process and the tea's essential oils. The characteristic brown colour and 'bite', meanwhile, come from the tannins present in tea leaves.

About 77 per cent of all the tea drunk in Britain in 1987 was made with tea-bags. In 1970 this figure was only 2 per cent. A typical tea-bag may contain up to twenty different sorts of tea, mostly small-leaved from African plantations. The small size of the tea particles inside the bag is very important. Since the juice from the tea-leaves dries on the outside of the leaves when they are rolled (using the CTC method), the finer the particles the more juice exposed. This means that an acceptable cup of tea can be made very quickly using tea-bags.

Colouring is never added to tea or tea-bags — there is in fact so much natural colouring matter in tea leaves (in the form of tannins), that the dregs from the tea pot could be emptied into a hot bath and the bather emerge with a slight tan!

The tea-bags themselves must be made of a very special paper that keeps its strength when wet. Many bags are made from viscose which will often have been made from the pulped timber of various hard and softwoods.

OTHER WAYS OF TAKING TEA The Russians who are the fourth largest producers of tea in the world drink their tea in a very distinctive way, using a special tea-urn called a samovar. The custom arose after a regular tea trade had been established between China and Russia in 1696. Finely powdered tea was pressed into large blocks for transportation and broken off in pieces for use. Today, the Russians drink their tea, brewed in the samovar, either with jam or by sucking it through a sugar lump. In Tibet, the home of China tea, the locals take their tea with the addition of rancid yak's butter!

PLANT FOODS FOR DIFFERENT PEOPLES

It is not just Western foods and drinks that come from plants. Everywhere in the world plants sustain human societies, but they are generally used directly, before any commercial processing. The famines that still occur with distressing frequency in tropical regions are caused not because of lack of agricultural skills but often because of the effects of climatic change, promoted or worsened by environmental mismanagement and by social strife. Under-nourishment in areas of tropical luxuriance is generally due to the disruption of traditional agricultural patterns by interference from outside.

Most traditional peoples who support themselves by various combinations of hunting, fishing or gathering from the wild and by gardening, eat far more plant than animal food. The diets of such peoples are in fact far more varied than is often acknowledged, and generally well balanced: high in fibre, with little alcohol and no salt or concentrated sugar except occasionally in the form of honey.

Much of what is eaten is fresh or in season and without the highly processed and excessively sweet foods that exacerbate or sometimes cause our own diseases, both physical and psychological health is very good.

Plants for cooler climates

Even the frozen Arctic offers plant foods for people. Eskimo groups have traditionally included a good deal of plant food in their diets, mostly from lichens, seaweed and different berries, as well as flower blossoms and grasses. Some groups eat around thirty different plants to supplement meals of seal blubber and oil, fish eggs, caribou meat, the organs of large sea animals and whale skin, rich in vitamin C.

With the melting of the winter snows, various berries including crowberries (*Empetrum nigrum*), bilberries (*Vaccinium myrtillus*) and cloudberries (*Rubus chamaemorus*), widespread across Inuit occupied areas of the Arctic, are picked and eaten. In the western Yukon-

Innuit girl gathering berries from the tundra

Kuskokwim delta region, medicinal teas are made from sourdock (*Rumex arcticus*), fireweed (*Epilobium angustifolium*) and the plant known commonly as labrador tea (*Ledum palustre*).

Plants for arid lands

In some of the driest regions of the world nomadic peoples like the Kalahari Bushmen and the Hadza of Tanzania eat a wide selection of fruit, seeds, berries, roots and important water-storing bulbs, as well as a variety of animal and bird meat when available. In the Kalahari, the nut of the manketti tree (*Ricinodendron rautanenii*) provides a staple source of protein for some Bushmen groups, especially during 'dry' years and the tsamma or karkoer (*Citrullus lanatus*), a water-storing melon provides a source of precious moisture. The fruit contains in fact more than 95 per cent water for many months after cutting and is also made into a stew by local people. The seeds, meanwhile, are ground to make an oily, nutritious flour. Despite their bleak surroundings, the daily calorie intake of the Bushmen (around 2400) is higher than that of the average European adult.

In Australia, Aborigines traditionally ate a varied range of plant and animal foods which formed a healthy and well-balanced diet. Today, many of them, resettled around towns, have diets high in processed foods, and as a result previously unknown problems such as diabetes, heart disease and obesity are increasingly common. The Bunaba people of Northern Australia,

Apple-berry (Billardiera scandens*) and pigface fruits* (Carpobrotus spp.) *are eaten by Aborigine groups*

for example, would have eaten around fifty different 'bush' foods regularly, from a total of about 200 locally available. The plants once commonly eaten by them and other Aborigine groups include succulent red or purple pigface fruits (*Carpobrotus* spp.), which are said to taste like salty strawberries; the sausage-shaped fruits of the common apple-berry (*Billardiera scandens*); *Grevillea* flowers, with a rich store of sweet nectar; seeds of the edible wattles, such as *Acacia coriacea*; bush yams and 'potatoes'; the seeds, roots and leaves of the pigweed plant (*Portulaca oleracea*); boab fruits (*Adansonia gregorii*) as well as the tree's succulent shoots and roots, and the billygoat plum (*Terminalia ferdinandiana*). This small green fruit which resembles an unripe gooseberry contains fifty times more vitamin C than an orange and was believed until recently to be richer in this important vitamin than any other fruit in the world.

Altogether, one third of our planet's land surface is arid or semi-arid, but every year this area increases by some 12 million hectares. It is estimated that almost 80 million people live in areas threatened by encroaching desert. This alarming transformation of land that once supported nutritious vegetation, caused principally through over-grazing, excessive human migrations and deforestation, has focused much botanical attention on plants that can survive in arid conditions.

The yeheb (*Cordeauxia edulis*), an evergreen shrub from Ethiopia's Ogaden region is one such plant. When conditions are too arid for other crops to grow, it will yield nutritious sweet-flavoured nuts.

The carob tree (*Ceratonia siliqua*) is now famous for the commercial use its fruit are put to as a chocolate substitute, but it cannot survive in extremely arid lands. In the Horn of Africa an endangered cousin of the carob (*C. oreothauma*), found now in only two locations in the wild, has the all-important capacity to grow in drought-ridden areas, and it is the current subject of a breeding programme aimed at introducing drought resistance into its better known relative.

Sadly, native plants are now in danger in many parts of the world from a combination of factors causing the destruction of natural habitats. Botanists everywhere are racing to document at least, if not to save, as many species as possible before they disappear. At the moment, no one knows exactly what is being lost nor what the potential of each plant species is to man, and this situation will not be helped as long as those best qualified to know, namely the indigenous peoples who have observed and learnt to use the plants around them continue to be moved away from their traditional lands.

Chicle tree (Manilkara zapota)

SWEETS FROM THE RAINFOREST

Chewing gum

The Mayan peoples of Guatemala were chewing gum from the chicle tree (*Manilkara zapota*) long before its milky latex was processed and sweetened for the first commercial use in the late nineteenth century. Today chicleros still tap wild trees which grow in the tropical rainforests of South and Central America for their latex, which is moulded into blocks before processing into sticks of chewing gum. Latex from other trees native to Central America is also used, and jelutong tapped from the South East Asian rainforest trees, *Dyera costulata* and *D. lowii*.

Many different peoples have chewed substances exuded from the bark of trees. For centuries the Greeks have chewed mastic gum — the resin obtained from the small shrub (*Pistacia lentiscus*) native to Greece and Turkey. In North America, the Indians of New England taught colonists to chew the gum-like resin exuded from spruce trees when their bark was cut.

A bar of chocolate

Chocolate comes to us courtesy of the South American rainforest. Throughout the tropical world cocoa plantations rely for their vigour and disease resistance on genes from their wild and semi-wild Amazonian relatives. At their most diverse along the eastern jungle-covered fringes of the Andes, the natural home of cocoa (*Theobroma cacao*) is threatened now as never before by oil exploration, human settlement, roads and airstrips.

The greatest area of modern cultivation is in West Africa, but the cocoa tree was being cultivated in Central America at least 2000 years before Europeans arrived. Cacahual, the pod-shaped fruits, were according to legend of divine origin, and part of the diet of the plumed serpent god Quetzalcoatl. It was this association that gave rise to the genus name *Theobroma* established by Linnaeus, meaning literally 'food of the gods'. Most famous as the essential ingredient of the drink (flavoured with vanilla, peppers and other spices) given to Cortés by Montezuma in 1519, cocoa beans were also used as currency by the Aztecs, who paid their taxes with them until 1887.

Cocoa beans contain the caffeine alkaloid theobromine, which is a mild stimulant. The chocolate drunk by the Central Americans, however, gained a reputation as an aphrodisiac, and Casanova is said to have preferred the substance to champagne as an inducement to romance. Chocolate is also said to contain the chemical phenylethylamine, a natural amphetamine found in the human brain, which induces a feeling of euphoria. The level of this chemical it is claimed, decreases when one is love-sick — explaining the craving to eat chocolate as a solace at such times!

SPICES

Spices from the tropics earn around $144 million each year in world trade. The lure of cloves, cinnamon, cardamoms, allspice and, most importantly, pepper — the only spice which could make decaying or heavily salted meat edible — drew Columbus and later medieval merchants to 'discover' for themselves the rainforested regions of the earth. For Europeans, North America was a 'by-product' of the maritime search for pepper, whilst at home Venice had become rich and beautiful by the sixteenth century from the profits of the pepper trade. When parts of the warship Mary Rose were recently lifted from the sea-bed in Portsmouth Harbour where they had lain since 1545, each sailor was found to have carried with him a small bag of peppercorns to spice his food.

Foods from the forest

Nowhere is this situation more desperate than in the world's tropical rainforests. Every minute we cut down or burn 100 acres (40 hectares) of the most diverse ecosystem on earth. Over half of all the world's plants, less than 1 per cent of which have been examined for their potential uses, are found in these forests. They give medicines, fibres and precious raw materials that help support us and enrich our lives each day. The tropical rainforests also offer us an enormous range of foods that liven up our diets, including chewing gum and chocolate, fruits and vegetables, nuts and spices.

But the forests also provide millions of people who live in and around them with their staple foods, chiefly root crops such as cassava, yams, cocoyams, sweet potatoes and taro. For some 500 million people, manioc or cassava (*Manihot esculenta*) is the single most important food, eaten at almost every meal. The tubers which form from the plant's adventitious roots are high in starch (carbohydrate) with low levels of protein, but processed in a number of different

---

THE MACHIGUENGA — A RAINFOREST PEOPLE

Sweet potatoes (Ipomoea batatas)

For the Machiguenga, a group of between 5 to 7000 Amazonian Indians who live in the rainforest of South East Peru, sekatsi as their manioc tubers are known have a special importance. They are believed to be presents from the moon who came down to earth in human form bearing both tubers and manioc plants. These were the first crop plants to be given to the people, with the instructions that they should be grown with care in their gardens, and treated well.

Though manioc, served with small amounts of meat or fish, is the Machiguenga's staple food, they recognize many different varieties. Diversity, as reflected by the other vegetables and fruits they eat, is indeed a horticultural principle. Including plants useful for household purposes, medicinal and fibre plants (including cotton), the Machiguenga grow around eighty different cultigens in their gardens. About thirty of these are to be found in a typical house garden at any one time — the most commonly planted of the food crops including maize, pineapple, sugar cane, yams and cocoyams. Of the rest, crops such as pigeon peas, peppers, taro, beans, avocado, squashes, soursop, peanuts, guava, plantains, bananas and mangos with their different tastes and textures, make interesting additions to their diet.

Like other traditional rainforest people, the Machiguenga are skilful and knowledgeable horticulturalists who for thousands of years have met their needs without irrevocably damaging their environment.

The people of the wet humid forests of Papua New Guinea have managed their forests in much the same way. For them, sweet potatoes (*Ipomoea batatas*) and yams (*Dioscorea* spp) are particularly important. Whilst sweet potatoes, the staple of many of the 700 or so different ethnolinguistic groups (and incidentally the second most important crop after rice in Japan) are prepared by the women, yams are grown and harvested in secret by men away from the women. At special yam festivals, prize tubers, some of them as long as 2 metres, are exhibited and decorated with faces. Reflecting the status of the grower within the community, these ceremonial yams are the focus of ritual exchanges and given as gifts at the end of the festivities to women. Of the huge number of yam species in existence many appear to have been domesticated independently in the Old and New Worlds. Some yam species have become extremely important to Western medicine. Diosgenin from several species of Mexican yam enabled the contraceptive pill to be produced, and is the starting material for many steroidal drugs today.

Andean woman with oca tubers, a relative of the potato

ways according to variety, they are a major source of calories. Some maniocs known as 'bitter' varieties (since they contain poisonous cyanide compounds which must be removed before the tubers can be eaten) are processed into various kinds of flour. Farinha in Brazil and gari in Africa are used to make flat breads and other starchy foods. Delicious sauces, such as casareep traditional in the north-western Amazon, and fermented drinks, are also important uses. In the West we eat rounded pellets of manioc starch in the form of tapioca pudding.

Foods from the mountains

Moving from some of the wettest regions on earth to some of the highest, another tuber, the potato, has risen from small beginnings to become the fourth most important crop in the world.

The famous staple food of the Andes, revered by the Incas, and grown today at altitudes too high for most other crops (up to 4300 metres), the potato is now the world's most important non-cereal food plant, accounting for over half of the annual harvest of starchy roots and tubers.

Each time we eat a packet of crisps, tuck into french fries, or simply peel a potato, we have Andean farmers to thank. Before 6000 BC the first wild potatoes were being collected from the high plateau, about 3600 metres above sea-level, that stretches between Cusco and Lake

Titicaca. Over thousands of years potato agriculture grew and flourished.

Of the eight different species of potato in existence, Andean farmers recognize 3 to 5000 different varieties. Some farmers today may grow up to 45 varieties in their tiny fields that cling to the mountain sides. Some amusing names are used to distinguish them: 'cats nose' (Michipasinghan) for a long, flat potato, or 'potato makes young bride weep' (Lumchipamundana) for one that is knobbly and hard to peel are just two of the thousand or so names for potatoes found in the Quechua language. The shape and colour of these tubers varies enormously, from yellow, round and twisted to purple, long and straight! Eighty per cent of all the potatoes grown in the USA and Canada, however, stem from just six varieties of the only species eaten in the West; *Solanum tuberosum.*

Scientists at Peru's International Potato Centre are working to save as many of the often small but genetically valuable Andean potatoes from extinction as they can, as local farmers are increasingly encouraged to plant modern higher-yielding kinds. A major part of the work at the Potato Centre has been helping adapt the potato to cultivation in the humid lower zones of the tropics where it has traditionally languished — drawing on the Centre's genetic bank of South American varieties.

Few single foods have as much nutritional value as the potato. Nutritionists rate the quality of its protein higher than that of the soyabean (one of the world's principal protein crops), and one single potato can supply half the daily vitamin C required by an adult. The potato is also 99.9 per cent fat free! One hectare's yield of potatoes gives almost as much food as two hectares of grain. With this prolific harvest it has been estimated that the average annual potato crop (291 million tonnes) could cover a four-lane super-highway circling the world six times!

Grown in 130 of the world's 167 independent countries, Russia and Poland grow most of the world supply. Holland has one quarter of all its arable land given over to potato production — a crop more lucrative even than tulips! Despite the fact that half the world's annual potato crop is being fed to livestock, potatoes still feed millions of us every day. The Americans eat over 50 per cent of their 16 million tonne potato harvest in the form of french fries and crisps. Europeans also have a special regard for the potato. The Irish respect it more than any other food in the wake of the horrific famine in 1845 that killed a million people and caused many others to emigrate to the New World. Belgium, meanwhile, boasts the world's only potato museum, and in France those judged to have helped promote and honour the potato are awarded membership of the Académie Parmentier, Grand Ordre du Noble Tubercule — an association of chefs, gourmets and restauranteurs who value more than many, the nutritious pomme de terre, the 'apple of the earth'.

FUTURE FOODS

The modern supermarket shelf is deceptive in the variety of foods it tempts us with. When the wrappers are removed, we find that most of these foods are derived from just a handful of plant species. Nearly three-quarters of the world's food supply is now precariously based on just seven major crops: wheat, rice, maize, potatoes, barley, cassava and sorghum. In fact we cultivate fewer species in the West than the farmers of neolithic times.

The price we have paid for huge increases in the productivity of our major crops over the past fifty years is a dangerous uniformity and under-utilization of additional or alternative food plants. Hundreds of neglected species, most of which have a long history of use by local peoples (who have a sophisticated understanding of their environment) could provide a diverse range of nourishing foods. As they attract scientific attention, new examples are constantly emerging of plants that could help feed the world's 500 million starving people and provide a better diet for many others.

Mangosteen (Garcinia mangostana*)*

Winged bean (Psophocarpus tetragonolobus*)*

The winged bean (*Psophocarpus tetragonolobus*) from Papua New Guinea, enthusiastically described as a 'supermarket on a stalk' is a good example. Every part of the plant can be eaten, from the immature pod, considered a delicacy by native people long familiar with the plant, to the dried seeds which contain a high quality protein and quantities of oil. The vine-like plant contains more protein than potatoes or cassava, and, set to become the 'soyabean of the tropics' can match the nutritional value of this now ubiquitous temperate crop.

Though we may marvel at the colourful selection of temperate and tropical fruits on display in our shops, they represent only a tiny fraction of the array we could enjoy. Estimates indicate that around 2500 different fruits from the world's tropical forests — only fifteen of which already rank as commercial species — could enhance our diets. Local favourites such as the soursop (*Annona muricata*) from tropical America, and the mangosteen (*Garcinia mangostana*) from Malaysia, described by some as the most delicious fruit in the world, are still the rather novel supermarket partners of pineapples and plums.

Many plants from cooler regions could provide vital calories and protein for people living in countries with harsh climates. Several Andean crops, such as ulluco (*Ullucus tuberosus*), oca (*Oxalis tuberosa*), and aiñu (*Tropaeolum tuberosum*) which yield tubers high in carbohydrates, and tarwi (*Lupinus mutabilis*) grown for its highly proteinaceous seeds, are possible candidates.

Marine plants are already valuable food resources in some parts of the world. Rich in sugars and starches with some protein and sometimes vitamin A and C, thousands of tonnes of seaweed are collected each year by the Japanese for traditional dishes such as sushi and nori. Once quite widely eaten in northern Europe (the best known British examples being dulse from *Rhodymenia palmata* and laver bread from *Porphyra umbilicalis*, both red algae) the nearest most of us come to eating seaweed today is in ice-cream and other soft or semi-liquid processed foods, which use their gelling substances, especially agar and carrageenan to help produce thick smooth textures.

The tiny spirals of blue-green alga, *Spirulina*, each one only 0.25 millimetre long are a lesser known aquatic food, but one with great potential for the future. Growing naturally in floating mats on Lake Texcoco in Mexico and on small lakes near Africa's Lake Chad, *Spirulina* is easily digestible when dry and comprises 70 per cent protein. The Aztecs made it into nutritious biscuits, as do the people of Lake Chad today. The alga is now being produced commercially in Mexico and already sold for human use in pill form, may come to be used as protein supplement in many other foods.

Back on dry land scientists are making other proteinaceous foods from green leaves. Where food is plentiful we tend to take our nutritious green-leaved vegetables for granted, almost unaware of their importance in providing us with protein, fibre, calcium, iron, folic acid and important vitamins. In many urban areas of the tropics, where diets are especially poor, the intake of leaves and sometimes shoots and stems is more or less essential for peoples wellbeing. Supplies are often obtained from the leaves of plentiful root crops like cassava, sweet potatoes or even fruit trees like the pawpaw, yet with most of the world's food production directed to the developed world far less 'green food' is eaten in general than in the richer temperate countries.

A process known as green crop fractionation has recently been developed by scientists to derive a high quality protein food directly from temperate or tropical greenery. Plant leaves and stems (leguminous plants have been favourites for experimentation so far) are crushed and pulped to release their juices which contain the all-important proteins, sugars, salts, lipids and vitamins. Heated to a temperature of 70 to 80°F, the proteins coagulate and can be separated out. Dark green and cheese-like in appearance, this solid 'protein cake' could become significant in protein deficient diets around the world.

Yet before we rush to dispense pills and protein supplements it seems sensible, if not essential, that we recognize the value of the food plants we can eat now, if not grow ourselves directly. If we can only leave their natural habitats alone and listen to the people they have fed for centuries, a great array of future foods awaits us. It took Europe two hundred years to accept the potato and the 'wonder' soyabean was generally unknown outside Asia until a few decades ago. The world's rich store of fruits, vegetables, tubers, grains and pulses is there for us to make the most of if we choose to do so.

SOYA

Glycine max

The soyabean which originated in Manchuria has been eaten in the East for centuries. It overtook wheat and maize as the most important of America's cash crops in the 1970s and is still the world's primary protein plant. Soyabeans can be processed in a great number of ways to make different foods. Textured soya protein can be flavoured to make anything from dog food to hamburgers and both oil and 'milk' — very similar to that produced by cows — are also extracted from the bean.

In Japan, miso, a soup traditionally eaten at breakfast time is made from the fermented bean paste. Tofu made from soyabean milk is also eaten in soups or with rice, meat and vegetable dishes. Soy sauce is derived from the fermented beans.

House and home —
plants that protect us

For as long as houses have been built, plants, chiefly trees, have played a major part in their construction. Whatever the age or style of your house, wood is almost certainly supporting, lining or adorning some part of it.

As the most versatile of all building materials — combining flexibility with durability and strength — and often the most readily available, wood continues to be indispensable for housing us and furnishing our homes. In North America and Scandinavia, between 80 and 90 per cent of all family homes are made of wood, and in Britain houses with timber frames now account for an increasing proportion of all those being built.

Around the world some of the simplest but none the less effective dwellings have been based on a wooden frame. The North American Indian wigwam relied traditionally on a conical framework of poles to support a weather-proof covering of hides, and the felt-covered yurt used today by nomadic peoples of Mongolia and Afghanistan is constructed from a wooden lattice often cut from willow trees. Both homes have the advantage of being very strong yet are easy to dismantle and reassemble as and when required. Pygmy groups in Africa, who make new homes as they move to different parts of the rainforest, bend saplings into a 'beehive' dome and cover this with a thatch of tough broad leaves.

In Europe, timber-framed buildings such as the Medieval 'cruck' and 'post and truss' styles appear to have developed from simple rectangular ridge-tent constructions in which poles were bent together in the form of an inverted 'V'. Immense mixed forests blanketed the continent — conifers predominating in the northern and Alpine regions, whilst sturdy oaks prevailed in central Europe and at its western edge, in Britain. Above all other native trees, oak with its tremendous strength and resistance to decay was soon recognized as the most suitable wood for making the framework and supporting structures of substantial buildings.

In 'cruck' framed houses, massive timbers, often the matching halves of a single split tree trunk joined together by a ridge beam running the length of the house, took the roof load straight to the ground.

As an understanding of the principles of timber framing and basic forms of bracing evolved, buildings became more complex. The 'post and truss' framed house developed as a secondary box-like structure built around the 'cruck' but with roof supports constructed separately from the wall frame. Where ground space was restricted, the great strength of the oak timbers allowed cantilevered upper stories or protruding jetties to be built above a narrow first floor. The gaps between the supporting wall posts were filled with wattle and daub — flexible twigs chiefly hazel, plaited together and covered with a mixture of mud and straw, or lathes (thin strips of wood) and a coating of plaster.

By the sixteenth century small bricks were also being used for infilling, but it was not until the eighteenth century, with the scarcity of suitable oak timbers and the development of solid brick construction, that the use of the timber frame declined in Europe. The tradition was,

Timber is a very popular building material in Europe, as this Black Forest house (left) and French alpine chalet (right) show

however, continued in North America and Australia by European colonists who took advantage of the vast natural forests offering a rich supply of timber. Distinctive architectural styles arose — often characterized by the use of exterior weather-board cladding — relying generally in North America on native softwood species and in Australia on an impressive range of eucalypts.

Influenced by these building styles and with the development of standardized, prefabricated frames and panels, the timber-framed house has made a popular come-back in Europe. But wood was never entirely superceded as a building material in the conventional brick-built house and has remained indispensable for the construction of rafters, joists and roof trusses, as well as skirting boards, staircases, traditional door and window frames, and flooring. As much as half a tonne of timber is likely to be used in this way in a standard masonry-framed house.

The demand for wood and wooden finishes within our homes has continued to grow. Though fashion has played a part in its appeal, very practical considerations such as the excellent capacity of wood to insulate both heat and sound and, where timber-framing is concerned, its 'dry' construction are important attractions. Over 1500 gallons of water may end up in the walls of an average brick house, in the mortar holding the bricks in place, and the plaster that covers them — this can sometimes cause cracking if the house settles and dries too fast. In the modern timber-framed house, prefabricated units that are easily assembled then covered and reinforced with other wood-based sheets or panels replace supporting walls of masonry and plaster. The walls are in fact made of a number of layers, which surround and fill the frame (or 'stud'), and are likely to include — working from the surface of the inner wall outwards — plaster board, a vapour barrier, insulation material, sheathing (such as plywood or a wood-based panel), breather felt and external brick or timber cladding.

WOODS FOR MODERN TIMBER FRAMES AND JOINERY

Roughly 200 000 new houses are built in Britain each year, and with all of them using wood to a greater or lesser degree plus continual additions and embellishments to existing homes, our annual consumption of timber for these purposes runs into millions of cubic metres.

Oak, in demand for centuries for the construction of ships and all manner of buildings as well as interior fittings and furniture is now a luxury timber, too scarce and expensive to be used on a large commercial scale for housing. Instead, most timber used today for framing and general construction work in Europe comes from softwood tree species chosen for their

Some early European churches used split oak logs set vertically into the ground and held together with wooden 'tongues' to form solid timber walls. Spaced at intervals on wooden bases or sills and braced by a lean-to formation, these trunks developed architecturally into vertical posts or staves able to support complex many tiered roofs.

Twelfth century European stave church

straight grain, low production cost and regular availability. Most of these trees are grown in large managed forest plantations. Scandinavian and other northern European countries have traditionally used their own supplies of spruce (*Picea*), pine (*Pinus*), birch (*Betula*) and fir (*Abies*) species. North American forests, especially those of British Columbia provide a large proportion of the softwood timbers used elsewhere in the world, and immense forests cover the north-western seaboard states. The trees grown in North America include Douglas fir (*Pseudotsuga menziesii*), sitka spruce (*Picea sitchensis*), lodgepole pine (*Pinus contorta*) and western hemlock (*Tsuga heterophylla*).

Balsa (Ochroma lagopus) *is a light, soft wood but it is technically a hardwood*

SOFTWOOD OR HARDWOOD

The terms 'softwood' and 'hardwood' can be somewhat confusing, since some softwood timbers — produced by gymnosperms (conifers or cone-bearing trees that are mostly evergreen) are physically harder than some of those produced by hardwoods (broadleaved, dicotyledonous trees, characterized by seeds that are contained in an enclosed case or ovary). Balsa (*Ochroma lagopus*) for example, one of the softest and lightest woods in the world, is technically a hardwood.

The two kinds of wood are distinguished anatomically in that they have different types of water conducting cells and supporting tissues.

Scots pine (Pinus sylvestris), *the only true pine indigenous to the British Isles grows up to 30 metres high and its trunk reaches 1 metre in diameter. The sapwood is creamy white to yellow, the heartwood yellowish or reddish brown.*

Softwood construction and joinery timbers

The pines

There are about ninety true pine species belonging to the genus *Pinus*. The pines are distinguished in general by their long slender needles and hanging cones, made up of overlapping woody scales. Their timber is divided in the trade into 'soft' and 'hard'. Hard pine is usually darker in colour, hardier, heavier, stronger and more resinous than soft pine. Whilst a large number of pines are native to North America — from *P. palustris* which grows in the Gulf of Mexico to *P. contorta* in the Yukon Territory — Scots pine is the only pine species native to the British Isles.

The spruces

The spruces are amongst the lightest of timbers used for general construction and joinery work today. Thirty-four species, characterized by needles that grow in spirals, belong to the genus *Picea*. They produce timber of a creamy white to yellowish brown colour with a straight grain that is generally easy to work. Some species have a very fine lustrous texture — notably western white spruce which is one of the most widely distributed softwoods in Canada. In Europe, Norway spruce, also sometimes classified as white or silver fir, is the most important species. Young trees are used as Christmas trees in Britain. Sitka spruce is also widely planted throughout the UK.

Norway spruce (Picea abies)

DOUGLAS FIR

Pseudotsuga menziesii

Much used for general construction work in its native North America and traditionally imported into Europe from Canada, Douglas fir is now widely planted in the British Isles and other European countries. The tree, which produces one of the strongest and stiffest softwood timbers, can reach heights of over 90 metres, with a diameter of 4.5 metres. The current flagstaff at Kew Gardens (69 metres high) is made from a single trunk of Douglas fir — presented by the Government of British Columbia. This massive tree was about 370 years old when it was felled.

WESTERN HEMLOCK

Tsuga heterophylla

Western hemlock which is native to the coastal regions of western North America, is one of ten hemlock species. Half of all the Western hemlock grown comes from large managed plantations in Canada's British Columbia. It is the largest of the hemlock species, growing up to 60 metres in height. Its light, uniform, knot-free timber, referred to as 'clear grade' is highly prized and is achieved by the very close planting of trees so that the lower branches, deprived of light, die off, leaving no knots. The strong, pale brown wood which does not splinter is one of the most valuable of North American timbers.

CALIFORNIAN REDWOOD

Sequoia sempervirens

The most remarkable and perhaps best known of native North American trees are the sequoias, including the Californian redwood (*Sequoia sempervirens*) and the giant sequoia (*Sequoiadendron giganteum*), also called Wellingtonia. Whilst the giant sequoias are famous for being the largest and amongst the oldest trees in existence (some are over 3000 years old, with a height of over 80 metres and a girth of over 24 metres), their wood is soft and brittle and of little commercial value. The Californian redwood, on the other hand, the world's tallest tree, has been extensively exploited recently for shingles and exterior cladding for buildings, as well as the construction of wooden pipes, tanks, vats, coffins and posts, for which the great durability of its rich reddish-brown timber makes it naturally well-suited. The timber has also been used for plywood, and its thick soft bark made into fibreboard.

The trees, once abundant, are now restricted to a narrow tract of land extending from southern Oregon to the Monterey Bay on the Californian coast, and a few stands on the western slopes of the Sierra Nevada mountains.

Douglas fir flagstaff at Kew Gardens

Western hemlock (Tsuga heterophylla)

Californian redwoods (Sequoia sempervirens)

About 90 per cent of Canada's harvest of these trees is used for timber-frame construction. Other widely used general joinery timbers include, from European forests, Norway spruce (*Picea abies*), Scots pine (*Pinus sylvestris*), and silver fir (*Abies alba*), and from North America a large number of pines, amongst them 'Southern' pine (which includes *Pinus palustris* and *P. echinata*) and Virginia pine, *P. virginiana*.

Logging in Canada

WHICH WOOD? TIMBER TERMINOLOGY

As do-it-yourself enthusiasts will know, the labelling of joinery timber is not usually specific as to the species used. Instead some very general terms are employed by the timber trade making it difficult for all but those growing the trees to identify them. 'S.P.F' for example, is used to label a softwood timber from any of the commonly grown spruce, pine or fir species, whilst 'Hem-Fir' generally refers to Western hemlock or any of the broadly grouped 'fir' species, but often amabilis fir. 'White deal' is usually Norway spruce or silver fir. European redwood or 'red deal' are terms for Scots pine often grown in Poland, Russia and Finland. These much used labels demonstrate the vagueness that has characterized the identification of a good number of the tree species used by the timber trade. The terms 'pine' and 'fir' for example are often applied very generally and botanically incorrectly to a range of coniferous trees. The situation has been complicated by the use of different names in different areas for the same species. A case in point is Douglas fir (*Pseudotsuga menziesii*) which has also been called Oregon pine and British Columbian pine. The tree is in fact neither a true fir nor a pine species, but belongs to a separate genus (*Pseudotsuga*). The fact that much of the timber destined for building work is processed in the country of origin and enters the world market as relatively anonymous sawnwood makes identification by the layman almost impossible. The rules and regulations governing the felling, labelling and sale of commercial timber world-wide are certainly very complex and have not helped the situation.

For timber-frame construction the naming of the tree species used is not regarded as necessary for general purposes and common trade names will often not distinguish between timbers from quite unrelated genera or places of origin.

The need for clarification is most urgent in the case of tropical hardwoods. Logging in many tropical regions is still being carried out in a highly haphazard and indiscriminate way with whole areas, often small islands in Malaysia and Indonesia, being entirely denuded of their trees. The introduction of machines that can reduce a large hardwood forest tree, and practically anything else growing around it to wood chips in a matter of minutes has exacerbated the situation and many producing countries, which export a proportion of their timber

PANEL PRODUCTS FROM TREES

Timber, our oldest construction material, is still the most versatile. New building materials which incorporate it and use it in new ways are constantly appearing. A relatively recent introduction is 'glulam', glue-laminated timber which comprises boards made up of small cross-sections of wood glued together in layers under pressure so that the grain of adjacent boards is parallel. Large pieces of any size or shape can be made. An important feature of the glulam is that it can take greater stresses than solid timber boards.

Glulam is just one of an enormous range of products available for use in house construction and interior work, made by sticking or pressing strips, veneers, chips, flakes or fibres of wood together in different combinations.

Plywood being made in Nigeria

PLYWOOD

Perhaps the best known panel product is plywood, which is made by sticking thin sheets or veneers of wood together in layers so that the grains are at right angles to one another, producing a very strong, inflexible board. The Egyptians made crude forms of plywood by bonding wood veneers together with glues made from animal bones and blood albumin. Today standard plywood veneers are produced by a peeling machine which peels logs rather like a pencil sharpener. The finished product is available in various strengths and thicknesses and with different specifications according to use.

Currently many softwoods such as Douglas fir and Southern pine from the USA and Canada, and tropical or temperate hardwoods — including birch from European forests — are used in different combinations to make plywood. Over 40 per cent of the plywood used in Britain, the second largest importer of plywood in the world, is made from tropical hardwoods including teak and rosewood, mainly from South East Asia.

CHIPBOARD

Since its development some 40 years ago chipboard has been steadily superseding blockboard and laminboard for many uses. Made chiefly from chipped coniferous wood residues (mostly Scots pine and Norway spruce), including forest trimmings and shavings from board mills, chipboard manufacture is much less demanding in terms of raw materials and skilled labour. Wood shavings are dried and blended with synthetic resins then pressed into mats, in a highly automated process.

It is not just trees that are used for making panels. Flax shive, (the waste woody matter separated from flax fibres during processing) and bagasse (the fibrous residue left after sugar has been pressed from sugar cane stems) are also turned into chipboards.

FIBREBOARDS

Many different types of fibreboard are in use today including softboard, medium board and hardboard. The first was produced in the late nineteenth century and contained large amounts of pulped newspaper which made the overall density very low. Later, ground wood pulp was used, and during the 1920s and 30s with the development of further techniques which could break solid wood down into fibres and then reconstitute them using pressure and heat, a much stronger and more durable panel was made: hardboard. To make these fibreboards, wood chips are softened by pre-heating in steam and then ground between two discs. Water is added to form a slurry and a number of other ingredients which may include rosin, drying oils, preservatives and fire resistant chemicals are mixed in to make different specifications. The water is removed by gravity, suction and special rollers to produce a mat in which the fibres are interlocked or felted.

Most of Britain's fibreboard is imported and is made from a mixture of hardwoods and conifers. Brazil and Australia are both big producers, with eucalyptus species (and especially *Eucalyptus saligna* in Brazil) being of major importance for hardboard panels.

in the form of plywood consignments, may be unaware of the tree species involved.

Though hardwoods are used for general construction work in tropical countries and for specialist building work and interior fittings and furniture elsewhere, these logging practices mean that hardwoods can quite easily be used for any number of prefabricated panel products that are now common features of temperate timber or masonry framed buildings. A new working surface in the kitchen or a panel covering a door could have begun life as a rare rainforest giant on an Indonesian island.

In the face of this obviously unsatisfactory position and with the world's rainforests being felled, burnt or flooded at the rate of 100 acres (40 ha) every minute, concerned individuals and environmental groups have called for urgent new rules to be enforced concerning the selective logging and replanting of tropical hardwoods and are pressing for an internationally agreed labelling system to be adopted showing the country and concession of origin. The Timber Trade Federation has agreed in principle to a code of conduct aimed at ensuring that only tropical timber from sustainably managed sources is allowed into Britain. The National Association of Retail Furnishers have endorsed this code and helped Friends of the Earth produce the Good Wood Guide, listing retailers stocking timber and wood products made from trees that are part of sustainably managed forests. The Guide lists only three tropical timbers that are produced on a sustainable basis; teak, greenheart and rubberwood.

House in New Hampshire, USA

House in Queensland, Australia

WOOD VERSUS THE ELEMENTS

Whilst timber cladding, in the form of sawn planks attached to the outside of timber framed walls is becoming more popular in Britain for long-term weather proofing and for decorative appeal, the USA, Canada and northern and Alpine regions of Europe have a long history of the use of timber weather boarding as an integral part of house structure and design.

Living in a log cabin or wooden alpine chalet makes sense where winter temperatures can fall dramatically. Timber cladding, particularly that with an internal air cavity, has far greater insulation properties than brick or concrete and will hold precious heat much more effectively. Ninety-eight per cent of Norway's low-rise housing is made of timber. The majority of Americans live in wooden houses too. Where the climate is harsh, timber which grows slowly acquires qualities of strength and density that make it an ideal construction material. In Europe and North America today the trees most commonly used for cladding include Douglas fir, Western hemlock, spruce, pine and larch species, Western red cedar and sequoia species. New techniques of pressure impregnation with chemicals have meant that timber no longer needs to be planed and treated with oil based paints to protect it from the weather and insects.

Wooden shingles, small rectangular pieces of wood produced by sawing against the grain have also been used for cladding and roofing and are an impressive feature of some traditional eastern European buildings. In America, shingles are often made of Western red cedar, though oak and Californian redwood have also been used. Local availability will determine to a great extent which woods are utilized. In Australia, eucalyptus has been the traditional building material for internal and external work and different species have been introduced to many other countries with harsh or dry climates. Both these and radiata pine — originally from California — are now major building materials in South Africa and Chile as well as New Zealand and Australia. In south-east Brazil and parts of Paraguay and Argentina, Parana pine produces wood for all kinds of construction and joinery work, and is also available in Britain.

Many American and Antipodean homes are complemented by timber decking, raised wooden platforms generally at first floor level which provide an open leisure area adjoining the house.

ON THE INSIDE

The use of hardwoods in our homes

In their countries of origin, tropical hardwoods (particularly those with special properties) have been very widely used for building work of all kinds. Teak in India, Burma and Thailand, iroko and afrormosia in West Africa, and jarrah in Australia, for example, have all been favoured because of their resistance to attack by termites.

Today, however, efficient preservative treatments and modern building techniques mean that a greater range of timbers can be used. As we have seen, softwoods are currently predominant for structural work in Europe and have been used for window frames, doors and interior joinery for many years, since they are cheaper, generally easier to work, and retain paint more readily than most hardwoods. Recently, however, tropical hardwoods — especially those with high density and strength properties — have been used for heavy construction work in industrial and institutional buildings, often replacing steel or concrete for joists, beams, portals and lintels etc.

Hardwood staircase at Longleat House, Wiltshire, England

Tropical hardwoods, especially those needing little maintenance, are also being used for interior joinery to satisfy the demand for 'the natural timber look', and it is here and as furniture that we find many of them in our homes today. In Europe, the trade in fine hardwoods from the tropics began in the seventeenth century — for the wealthy an age of grand residences with sweeping staircases and rooms highly adorned with wooden panelling. Mahoganies from the West Indies and Central America were amongst the most sought-after and during the eighteenth century the trade increased, with fine timbers such as rosewood from South and Central America imported in large amounts. Teak from India and later Burma and Thailand was traded extensively during the nineteenth century and towards its close African mahogany began to be shipped to England along with other decorative woods for furniture and household fittings.

HARDWOODS

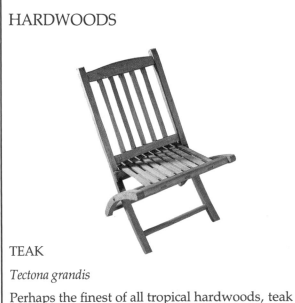

TEAK

Tectona grandis

Perhaps the finest of all tropical hardwoods, teak enjoys world-wide fame because of its strength, durability, stability in fluctuating atmospheres, the ease with which it can be worked and its beautiful appearance. It has been used for many purposes in Western households including all types of furniture, top-class joinery, panelling, doors, staircases, planking and veneers. Other important applications include boat and shipbuilding, dock and harbour work, piling, bridges, and sea defences. Teak is also valued for industrial purposes where a resistance to corrosive acid is required.

Tectona grandis is indigenous to Burma (the world's major exporter) and the greater part of the Indian peninsula and Thailand. Due to the great demand for timber, numerous plantations have been established within its natural range, as well as in Sri Lanka, parts of the West Indies, Central America and tropical Africa.

Teak trees have small white flowers and fleshy fruits. Their aromatic golden-yellow heart wood becomes brown when seasoned and is sometimes patterned with dark markings. The best Burma teak is typically straight-grained, though the grain of the teak from India is often wavy. Teak is unique in that it does not cause rust or corrosion when in contact with metal and its oils make it resistant to insect attack.

ROSEWOOD

Dalbergia spp.

Rosewood takes its name from the rose-scented timber produced by several tropical evergreen trees of the genus *Dalbergia*. These include Brazilian rosewood or blackwood, East Indian rosewood, used for cabinet work and veneers, and Honduras rosewood, a more ornamental timber used for musical instruments.

Familiar as a decorative wood for some 200 years, Brazilian rosewood has been so widely utilized that all easily accessible stands have now been logged and the timber has become scarce.

Brazilian rosewood trees are distinguished locally on the basis of the prevailing colour of their timber which may vary from chocolate or violet brown to a rich purple black, marked with irregular black streaks. The timber has a rather oily appearance and when worked releases a distinct though mild and fragrant odour. The main use of Brazilian rosewood today is veneer for furniture and cabinet making, though large amounts are used for musical instruments, knife handles, spirit levels and billiard cue butts.

Though the name rosewood is shared with trees of the *Dalbergia* genus, fragrant rosewood oil — highly prized by the perfumery trade — is extracted from trees belonging to the genus *Aniba* (of the Lauraceae family). The presence of linalool in *A. rosaeodora*, native to Amazonia, has led to the extensive exploitation of this species and the trees are now endangered in the wild.

MAHOGANY

Swietenia spp.

The name mahogany has been a favourite term for a number of tropical hardwoods. Today two 'true' mahogany groups are recognized, those describing several *Swietenia* species (the 'original' mahoganies) from Central and South America and the West Indies, and the related African *Khaya* species. Some 400 years after *Swietenia mahagoni* was first shipped to the UK and Europe, to become the most cherished cabinet wood in the world, this species has been so widely exploited that its significance is now largely historical. Despite its great value for furniture-making enormous numbers of smaller trees were used to fire the boilers of Caribbean sugar mills and steam trains, whilst many larger trees were felled for fence posts and railway sleepers. This indiscriminate use of the timber finally led the governments of Cuba, Haiti and the Dominican Republic to ban the export of mahogany logs in 1946, and today other Caribbean countries can only supply small amounts intermittently.

Another species, *Swietenia macrophylla*, has come to fill its position, but the demand for this timber has led to felling at smaller diameters and to the establishment of plantations in which growth is often rapid, but the wood less dense. *S. macrophylla* is distributed from South Mexico southward along the Atlantic slope of Central America from Belize to Panama; also in Colombia, Venezuela and parts of the upper reaches of the Amazon and its tributaries in Peru, Bolivia and Brazil. Thus a number of names are used for the timber according to the country of origin.

The mahoganies have large compound evergreen leaves, with greenish-yellow flower clusters and fruit capsules containing winged seeds. In the wild they can reach 45 metres in height with a diameter of 2 metres, above heavy buttress roots. The sapwood timber which is yellowish-white in colour is sharply demarcated from the heartwood, which is pinkish or salmon-coloured when freshly sawn and dries to a light reddish brown with a golden lustre. The grain often produces an attractive striped pattern and deposits of dark coloured gum in the pores are common. South and Central American mahoganies are still much in demand for high class furniture and cabinet making. Whilst 90 per cent of Brazil's exports go to the UK, Mexico, Belize and Nicaragua supply about 70 per cent of the mahogany imported into the USA.

After World War I both the quantity and range of tropical timbers imported from West Africa and Malaysia rose greatly and this expansion was repeated again after World War II. West African supplies were, however, soon eclipsed by the great growth of trade from South East Asia and more recently South America. Malaysia is currently the largest exporter of tropical timber in the world.

THE REAL PRICE OF A TROPICAL HARDWOOD DOOR

Buying a piece of tropical hardwood for a new door or window frame may be hard on the pocket, but it is often much harder on the rainforest from which it was taken. Every year around 5 million hectares of tropical rainforest is cut down in South America, Africa and South East Asia in order to extract timbers (chiefly for export to the developed world) to build, decorate and furnish our homes. Around sixty different tropical timbers are available for use in British homes today.

Ninety-five per cent of those tropical timbers used in Britain alone (the largest importer in Europe and amongst the largest of the consumers in the developed world) come from forests that are poorly managed or whose timber is being cut down on a non-sustainable basis. Since 50 to 90 per cent of all living things on earth are thought to live in the world's rainforests, scientists estimate that fifty species could be becoming extinct every day from the joint effects of logging, cattle ranching and other destructive practices. Unless this destruction stops and sustainable plantation production introduced we could well have extinguished one quarter of all the species on the planet in just ten years from now. In fifty years, rainforests could have entirely disappeared.

Temperate hardwoods are a cause for concern too, as the supply of large trees which have often taken hundreds of years to reach maturity is being rapidly used up, and many trees are severely damaged by the effects of acid rain. Since no standard system has yet been introduced by the timber trade for the classification of endangered or threatened species, Friends of the Earth are urging everyone to avoid wood which may have come from a threatened species by buying hardwoods only from shops which display the FOE 'Seal of Approval' sign. This means that the wood has come from a renewable source.

Furniture from plants

Every room in every house in Britain could quite easily contain furniture and household accessories — both modern and antique — made wholly or partially from tropical hardwoods. Mahogany, teak and rosewood are three that occur very frequently in solid and veneer forms in tables and chairs, wardrobes, shelving, cabinets, work tops, lamp stands, chopping boards, saucepan handles and toilet seats!

These items could also be made from temperate hardwoods, which have been used for centuries to furnish our homes. Whilst oak and walnut, two of the most important woods for traditional British furniture are too expensive now for wide-scale use, beech, ash, birch and sweet chestnut, in various combinations and often with other materials such as chipboard, canvas, steel and glass, are widely available today.

Other temperate woods often used by individual craftsmen, with a long history of use for decorative carving, veneer and inlay work include pear (*Pyrus communis*), which is a striking terracotta brown or reddish gold in colour; sycamore (*Acer pseudoplatanus*), a milky white wood that has a natural lustre and polishes to give an ivory-like surface; box (*Buxus sempervirens*), a fine even-textured wood, ranging in colour from pale yellow to bright orange; apple (*Malus*

TEMPERATE HARDWOODS

EUROPEAN BEECH

Fagus sylvatica

With its first fine veil of soft translucent green in springtime and leaves of copper gold in the autumn, the beech tree with its elegant grey trunk is one of Europe's most impressive native trees. Slim and conical in outline when young, older trees have a massive many-branched dome which spreads to form a thick overhead canopy. The dense shade produced in summer and the rain-drip from the leaves prevents the growth of weeds and the great quantity of leaves which fall from the beech each year adds humus to the soil.

In dense forests the beech tree can grow to a height of 45 metres, with a clear bole of around 15 metres, yielding a substantial quantity of straight-grained timber. The texture of the wood is fine and even but density and hardness vary according to the region of origin. When freshly cut the timber is a very pale brown, turning to pink or reddish brown on exposure to the air and a deeper colour after the steam treatment commonly given.

Whilst 'green' (before drying-out in the open air or by kiln) its strength is similar to that of oak, but after drying beech becomes 20 per cent stronger in bending strength, stiffness and sheer and considerably stronger in its resistance to the impact of heavy loads. There are innumerable applications of beech wood in the home and it is now the most commonly used native hardwood in Britain. Interior construction and joinery, mass-produced furniture (now superceding stripped pine), flooring and all sorts of domestic woodware use large quantities each year.

Since beech wood bends so well and can be turned so easily it is an ideal wood for making furniture, especially chairs. In the Chilterns, where beech chairs have been made for centuries by 'bodgers' (traditional chairmakers), beech woods are still managed to supply this ancient industry. Unfortunately much of our native beech woodland, especially in Southern England has now been felled and replaced with quick-growing conifers.

BIRCH

Betula spp.

Found throughout Europe from Lapland southwards to central Spain and from Ireland to Russia, birch is one of our most versatile hardwoods. It grows further north than any other broad-leaved tree and yet will tolerate great heat. In very cold areas the trees may never grow beyond the size of small shrubs, or will branch

Beech wood in Sussex, England

House in birch forest in New York State, USA

just above the ground, but in more equable climates they can reach heights of 18 to 20 metres with the trunks free of branches for some 9 metres. Two birch species out of sixty are recognized for commercial use in the timber trade: *Betula pendula*, the silver or white birch and *B. pubescens*, the common birch. Whilst the silver birch is distinguished by having rough silvery bark with drooping branches and smooth, short, pointed leaves, the common birch has a smoother, often reddish bark at its base with branches that are closer together and more horizontal. Its twigs and leaves are covered with soft hairs.

Birch timber is whitish to pale brown in colour with no distinct heartwood. It is fairly straight-grained and fine textured and when dry is similar to oak in terms of strength, and superior in stiffness. The biggest uses of birch timber are marquetry, veneers and strips for plywood and blockboard, dowelling rods, interior fittings and furniture.

Birch trees have long been connected with magic and the supernatural. It was considered lucky to hang birch branches over the doors of houses on midsummer's eve as the tree was believed to have protective powers and to avert the 'Evil Eye'. It is no coincidence therefore that the traditional 'witch's' broomstick was always depicted as being made of birch.

Mature oak in deciduous woodland

OAK

Quercus robur, Q. petraea

Of all the trees native to the forests of Europe the mighty oak has held a special fascination since the earliest times. The tree was sacred to the ancient Greeks and at about the same time to the Celts, who emerged from the Rhinelands of Central Europe. Mistletoe, the semi-parasitic shrub which will cling to the oak's branches under the right conditions was also regarded as sacred since it was thought to be the guardian of the tree. Both were the object of secret rites performed by the Druids, Celtic priests. In Norse mythology the oak was sacred to Thor, the God of Thunder and if struck by lightening, pieces of splintered wood were kept as protective charms. This king of British forest trees, which can take 300 years to reach maturity and at least eighty for the trunk to grow to 50 centimetres in diameter, became very early on the foremost construction material of Europe, and remained the most important furniture timber from the Middle Ages to the seventeenth century when it began to be displaced by walnut.

Not only very strong and durable, the natural formation of its branches allowed timber to be cut from the tree in curved shapes suitable for the cruck frames of medieval houses and the frame supports for ships. Oak timber had become so important to British shipbuilding by the sixteenth century that King Philip of Spain ordered the Armada to burn and destroy every oak that grew in the Forest of Dean — the source of much of the British Navy's timber at the time. Although the Armada was defeated, so many oaks had already been felled that extensive laws were passed by Queen Elizabeth I to protect them for future use, and extensive planting was subsequently carried out in the Royal Forests.

Despite centuries of use for iron smelting, tanning and construction purposes, much of southern England still had a large amount of woodland cover up to 1945. Between 1945 and 1985 over half of the remaining hardwood forest was clearfelled or severely degraded — an amount equal to that destroyed during the preceeding 400 years. Scarcity of good oak has now made it the most expensive of British timbers and it is therefore often supplemented by faster growing American and Japanese oak, although these are less attractive to carpenters.

Oak ecology

The oak, for so long part of the European landscape, fulfils an important ecological role. It is estimated that some 500 different species of insect alone depend to a greater or lesser extent on the foliage, the deeply furrowed bark or the trees' timber for their existence, thereby providing an important source of food for many birds. The tree has also been directly beneficial to man and his domestic animals as a source of food. In Saxon times poor people ate acorns during periods of famine and towards the end of the seventh century special laws were made relating to the feeding of pigs in woods, on acorns and beech nuts. These 'rites of pannage' as they are known today are still observed by the commoners of the New Forest.

Oaks are also beneficial to the soils they grow in (often heavy clays), in that they play a major part in draining them. A mature tree will draw up through its roots around 20 gallons of water a day — a natural water pump for soggy ground.

HMS Victory

Oak timbers — characteristics and uses

The oaks of Norse and Celtic legend classified today by the timber trade as 'European Oak' refer to just two of the 600 or so separate species in the genus *Quercus*: *Q. robur*, known also as the pedunculate oak and *Q. petraea*, the sessile or durmast oak. Both of these are also often called English, Polish, Slavonian oak, etc, by traders according to the country of origin, and are further classified on the basis of their wood structure as 'white'. Since there is little difference between the appearance of the woods of *Q. robur* and *Q. petraea*, (the sapwood in both being merely lighter in colour than the heartwood which is a yellowish

continued

brown), the species involved is not usually distinguished by the timber trade. Growing conditions do affect the character of the timber though and trade names chiefly mark these differences. Baltic countries for example, including northern Poland, export oak which is generally very tough and hard, but the rich black soil of south-east Poland produces Volhynian oak, a less dense but very even timber. The weight of the wood is likewise determined by the region of origin. All 'white' oak however is extremely strong and durable and noted for its ability to resist the passage of liquids.

For furniture and cabinet making Slavonian, Volhynian and Spessart oak (from Germany) are generally preferred in Britain whether in solid or veneer form, whilst the tougher English oak is often considered more suitable for construction purposes.

Oak timber has a long history of ecclesiastical use — for the interior fitting of churches and cathedrals — and for marine construction, the building of boats, docks and sea defences etc. Its water-tight timber has also made it very suitable for barrel making and it is still essential today for the storing of precious liquors such as whisky while they mature.

Quercus petraea, *the sessile oak*

sylvestris), generally a soft pink, and lime (*Tilia europaea*), pale creamy yellow in colour and relatively soft but dense.

Softwoods commonly used for furniture in Britain include a very wide range of pine species, not generally specified individually by manufacturers but currently most popular in the form of 'antique' pine — timber chemically treated to look old — and yew (*Taxus baccata*), Britain's densest softwood which has a rich red-brown timber. The pines are likely to be grown in Scandinavia and other European countries such as Finland and Yugoslavia. The more northerly the origin and the slower the growth of the tree, the better the timber is said to be, of a finer, denser quality.

Pine is very popular for furniture-making

We tend to take our furniture for granted, but design and availability have progressed only gradually over the centuries. Until Elizabethan times only people of high social rank in Britain would have possessed chairs, which were solid constructions made of a heavy frame of squared timber. Those of lesser rank had benches and stools; stools were in fact the forerunners of what was to become the first chair available to the cottager, the Windsor chair, distinguished by its turned, light-weight legs and back supports (not added until the sixteenth century) set into the seat. Many different woods have been used: beech, birch, apple or pear for the legs, elm or oak for the seats and hooped backs often of steam-bent yew.

Other plants have helped to make our seating arrangements more comfortable. Until recently many upholstered armchairs and sofas were stuffed with vegetable horsehair — also called 'crin vegetal' consisting of fibres obtained from the leaves of a dwarf palm, *Chamaerops humilis*. Found in North Africa and the Mediterranean region including Italy, Spain and southern France, this is one of only two palm species native to the Mediterranean region.

Another palm, from South America, helps us to keep our wooden furniture in good condition. The Brazilian wax palm (*Copernica prunifera*) is an important source of wax for furniture polish. Carnauba wax, which forms on the palm leaflets as they develop, constitutes between 5 and 20 per cent of some household furniture polishes, which generally have a base of paraffin wax and beeswax. The main attraction of carnauba is its hardness — only small quantities are needed to give good protection and an attractive shine to the surface being polished.

Plants have also been involved in the production of a number of wood 'finishes', used partly to seal the grain against the atmosphere and so limit the wood's tendency to warp and shrink, partly to protect its surface from becoming marked or dirty and also to enhance the beauty of the grain. Linseed oil for example, from the flax plant, is still a major ingredient in many oil-based varnishes. Valued for its 'drying' properties it has been widely used to give preliminary protective coatings to oak, mahogany and walnut surfaces. In the past, the oil was sometimes darkened by the immersion of alkanet roots (*Alkanna lehmannii*), which were a traditional source of deep red-brown dyes. The original base of the various types of French polish is shellac, the secretion of a parasitic insect which feeds on *Butea frondosa* trees in India and Thailand. Shellac is processed by heating and dissolving in industrial alcohol.

Whilst true French polish is being superseded by polyurethane and polyester finishes that are more resistant to scratching, heat and to staining from dampness, Japanese and Chinese lacquer-ware still depends on resin produced by the lacquer or varnish tree *Rhus verniciflua*. The use of lacquer on furniture and other household items had become an art form in China before the first century AD, but was perfected later in Japan. Up to 300 layers of filtered resin (which oozes from the bark of the trees when cut) are applied, each one rubbed with charcoal powder then polished after drying to give a final glass-like finish. Fine designs in gold leaf or rice paper are laid onto the surfaces and coated with additional layers of lacquer.

Brazilian wax palm (Copernica prunifera),
an important source of wax for furniture polish

Salix viminalis, *one of two willow species classed as osiers*

Basket weaving in Cyprus

Plant stems for furniture

The stems of plants belonging to two very different families — one native to the tropics and the other to temperate areas — have given us distinctive cane and wicker furniture.

WICKERWORK. Wickerwork is a very ancient craft. The Roman scholar Pliny recorded the Etruscans reclining upon wicker couches and the Romans themselves favoured wicker furniture. Then, as now, the material involved was willow. Under the general name of 'osiers' in Britain, the young supple twigs of several willow species have been made into a range of household items, chiefly baskets and chair-seating and backing. During World War II food and ammunition hampers made from willow were used by the British since they could be air-dropped without breaking. Commerical use of the basket willows or osiers for any other purpose at that time was banned.

Today only two species, *Salix viminalis* and *S. purpurea* are classed as osiers, though the French willow (*S. triandra*) has also been extensively used for wickerwork. Both are found in marshy areas throughout central and southern Europe and Asia and are often cultivated. Rather than being pollarded like some other willows to produce straight poles, the osiers are coppiced, that is cut down to ground level once a year so that a mass of long pliant stems or 'withies' is produced. Much in demand at one time for chair-seating and for a number of other purposes including fish traps, lobster pots and different kinds of basket, commercial withy growing is now only practised on a small scale in Britain, the craft having suffered greatly through the introduction of mass-produced furniture.

Climbing rattans in swamp forest

RATTAN

Calamus caesius

The stems of this moderate-sized rattan which grows high into the forest canopy, reach about 100 metres in length. *Calamus caesius* is widespread throughout the lowlands of Sabah — where it is known as 'Rotan Sega' — and elsewhere throughout Borneo, Sumatra, the Malay Peninsula, Southern Thailand (where it may have been introduced) and Palawan.

It is acknowledged to be the best quality cane of its size and class, and is considered ideal for all kinds of furniture work. Rattan is a very versatile material. While the larger diameter canes make stable frames for furniture, others may be split into fine strands for chair backs and seating, for basketry and decorative screens.

Rattan furniture

RATTAN CANES. Used for chairs and many other light but sturdy items of furniture, and usually described as 'cane', are rattans — the rope-like stems of various thorny climbing palms from the tropical forests of South and South East Asia. Rattans belong to a subfamily of the palms, the Calamoideae. Two genera supply the bulk of all rattans used in the furniture trade — *Calamus* (of which there are some 370 species) and *Daemonorops*. In all about 20 to 25 species provide the most sought-after, fine quality canes.

In South and South East Asia the rattan trade is extremely important with an export value of hundreds of millions of pounds each year. Rattans are in fact the main commercial product of South East Asia's tropical forests after timber. About half a million people make their living by harvesting and processing the stems and in rural areas particularly they provide a vital source of income. Furniture is the biggest end use of rattan canes, and many European, Japanese, American and Australian companies import the finished products or the canes to make them.

Mature rattans have been known to grow to a length of 186 metres, though 75 metres is more usual. They grow upward and into the forest canopy with the aid of backwardly curved spines attached to 'whips' borne at the end of the leaves or on leaf sheaths. The stems grow either singly or send out a number of suckers from the base of the plant. Once a stem has reached the maximum diameter for the species this does not increase with age but the stem grows longer, towards the source of light. To extract rattans from the forest, accessible stems are cut close to the ground and then dragged from the supporting branches. After removing the leaves and spines, the stems are cut into more manageable lengths of 3 to 5 metres. They are dried for about a month and gathered together in bundles of 100 stems or more from several different species for grading. After grading the rattans are washed, scoured with sand and then bleached in the sun or artifically. Depending on their diameter, rattans will be used to make both sturdy frames and decorative trimmings and facings for furniture and a great number of other items used locally. Thailand, China, the Philipines and Indonesia are all important exporters of furniture and furniture canes though from 1989 Indonesia will only be exporting ready-made articles.

About 150 000 tonnes of rattans are harvested each year, almost entirely from the wild, and many species are now threatened with extinction. The problem is especially acute in Sulawesi where several species thought to be new to science may become extinct before they have been studied. Whilst the need for a sustainable harvest of wild rattans is urgent, some plantations have been set up in areas of pre-cut or specially prepared forest in Sabah, the Philippines and Thailand. Trees of other species are left to support the rattans as they grow, and the canes can be harvested relatively easily after 7 to 10 years.

Bulrushes being harvested in Oxfordshire, England

Rushes are soaked to make them pliable then twisted before being woven

True bulrush
(Schoenoplectus lacustris)

RUSHES FOR CHAIRS AND FLOOR COVERINGS. Another traditional material used for making the seats of chairs in temperate areas is *Schoenoplectus lacustris*, the true bulrush. Though the craft has declined greatly in Britain, bulrush cutting is still carried out in some parts of the country, chiefly from the Great Ouse and Thames rivers, supplying a small number of businesses and individual rush weavers. Lengths of around 1.8 metres are usually harvested in late summer, the stems being cut on or just below the surface of the water, and carefully dried before use. *Schoenoplectus lacustris* is a freshwater rush, native to the swampy areas and waterways of Europe, North America, North Asia, Australia and some Pacific Islands. It has been used as the traditional rush for chair seating in Britain since the seventeenth century and in some parts of the world is still used to make matting. A number of other marsh plants, often simply described as 'rushes' are also used for seating and mat-making. Many of the firms still using rushes in Britain import them from Holland.

Juncus acutus, the traditional mat and basket rush of Italy has long been used to make the soft, padded covering for Italian wine bottles and receptacles for olive pulp before pressing, and is still a raw material for chairs and matting. *Juncus maritimus*, known simply as the sea rush or sparto, is similarly woven into floor coverings, especially in Spain and Morocco, whilst the rush know colloquially as Chinese sea grass (*Cyperus tegetiformis*) could be covering the floors of British kitchens. It is exported in large quantities from China and has a long traditional use there. In India *C. pangorei* or Calcutta mat rush is an important raw material for mats.

Plants on the floor

In many parts of Europe long before the days of carpeting or other mass-produced flooring materials, rushes were amongst the principal plants used to make simple coverings — either strewn on the floor or plaited or woven into mats. In the poorest homes these would have been renewed only once a year. The carpets and rugs that most of us possess today still depend on plants but in quite different ways.

JUTE

Corchorus spp.

To extract their fibres tall jute canes are cut by hand close to the ground and left lying in the fields for a day or two to allow the leaves to fall from the stems. The stems are then made up into bundles and steeped in water for several days so that their fibres become loosened. The farmers strip the fibre from the plants whilst standing in water, beating them with wooden flails and washing them to remove the remains of the outer bark. The fibre is finally dried in the sun, baled and then sorted, ready for export.

Dundee, the current centre of the UK jute processing industry once supplied the whole world with jute goods. Its importance developed in the early nineteenth century after a long history of flax and wool spinning and weaving. Dundee was at that time a thriving whaling port, and whale oil was used initially to help process the jute fibres. Today, mineral oils do the job.

Grown several thousand years ago in Asia, the leaves were used first as a vegetable and source of medicines, and only later were the fibres utilized for rope and cloth. Whilst 90 per cent of the jute processed in the UK now goes into high quality carpet backings, the fibres are also used extensively for sandbags, sacks, ropes, cables and twine and for the manufacture of roofing felt, damp courses, plaster board, caulking for the decks of ships, vehicle interior linings and the soles of 'rope' sandals.

LINOLEUM FROM PLANTS

Whilst jute has been the traditional material for backing lino, the bulk of this tough yet pliant floor covering comprises several other substances from plants. Lino takes its name from the Latin 'linum' for flax, and 'oleum' oil, since it was originally made from a base of linseed (flax) oil. This oil is still used today along with other commercial seed oils, large amounts of rosin (from various pine species) and plant gums such as agar and algin.

After thickening with oxygen, the oils are mixed with ground cork or wood, gums, resins and pigments, before pressing onto a jute backing. Whilst lino is still manufactured, plastic floor coverings have displaced its use in many areas. A number of these plastic coverings, however, are also based on polymers derived from seed oils.

Matting made from coconut fibre or coir

Jute harvest in Bangladesh

CARPET SIZES FROM PLANTS. Though some non-wool carpets are made of synthetic oil-based materials, a large number are made from viscose or similar fibres, derived from wood pulp. Many carpet fabrics further depend on plants to provide the gums and starches needed to make sizes. Huge quantities of sodium alginate from seaweeds such as *Macrocystis pyrifera* are used for sizing carpet fibres. Potato starch and guar gum are also widely utilized.

JUTE FOR FLOORS AND FURNITURE. There could be plant materials beneath the carpet too! The backing or scrim, a loosely woven cloth stuck to the underside of many tufted and woven rugs and carpets, is usually made of cotton or jute. Jute in particuar has been of very great use to flooring manufacturers for its exceptional strength and durability. It is used to back most Axminster and Wilton carpets, to keep them stable and given them a better 'handle'.

Jute is also used to back quality linoleum and carpet underlays made of foam rubber and to line carpet underfelts — which may themselves be made wholly or in part of jute fibres. The uses of jute cloth in the home, usually in the weave known as hessian, continue in the form of wall coverings, the lining of upholstery fabrics (furniture webbing), and bed spring coverings.

Jute fibre is separated from the stems of *Corchorus* species, annual plants which flourish in monsoon climates and grow to some 2.5 to 4.5 metres in height. About 90 per cent of the world's supply comes from Bangladesh and India, the rest from China, Brazil, Formosa, Burma, Nepal, Thailand and Japan.

Bangladesh derives half of all its foreign exchange earnings from jute, its most important export crop. There, as in India and Pakistan the crop is cultivated by thousands of small-scale farmers, each growing some 1 to 2 hectares annually. More than one million hectares, in fact, are given over to jute cultivation in these countries each year. After cotton more jute is consumed today than any other vegetable fibre, and around 5 million people gain all or part of their living from the cultivation of the plants, or the sale and handling of items made from them.

COCONUTS FOR DOORMATS. A different tropical plant fibre — coconut — could easily be the first thing that welcomes you as you cross the threshold of your home! Better known as coir, it has been the traditional fibre used to make door mats for many years. India is the chief producing country and large amounts are also exported from Indonesia and Sri Lanka.

Much in demand for its natural resilience, durability and resistance to wetness, coir is stripped from the outer husk and shell of the inner kernel of the coconut fruit. Also known as 'cocos fibre' it is classified as one of the hard industrial fibres which enter world markets and is imported in the form of raw fibre, spun yarn or ready made floor coverings.

In Portugal pigs are released every autumn into the extensive cork oak forests to feed on acorns. The owner of the animals is charged a fee for this privilege, calculated by the weight of each pig both before and after the six or eight weeks spent in the forest!

It has been estimated that approximately 200 000 tonnes of acorns are fed to pigs in Portugal alone — converted annually into some 18 000 tonnes of pork!

Cork oak forest (Quercus suber)

TREE BARK ON THE FLOOR. The bark of an evergreen oak tree (*Quercus suber*) native to southern Europe, has provided us with a very popular modern flooring material: cork. Composed entirely of dead, almost completely water-tight cells held together by a strong resinous substance, the great value of cork as a flooring and panelling material lies in its absorption of sound and vibration, its lightness, unique elasticity and resistance to the passage of liquids.

Cork is taken from the oak by making vertical and horizontal cuts into its bark with a special axe or curved saw and then prizing off large pieces. As long as neither the inner layers of bast material which carry the sap up through the tree, nor the cambium cells which produce new wood and regenerate new protective layers of cork bark are harmed, this operation can be performed without permanently injuring the tree. The bark will renew itself naturally, with fresh layers growing each year.

Large quantities of cork are currently being cut from carefully managed forests in Mediterranean countries, to provide us with tiles for floors and walls. For these purposes the cork is compressed so that the natural resins bond the individual granules together, leaving a firm but cushioned surface that is very durable. As thermal and acoustic insulation for buildings cork has lined floors, walls and ceilings in southern Europe since Medieval times — affording protection from the intense summer heat yet providing a surface always warm to the touch in winter.

Cork has also proved itself the ideal material for stoppering bottles as it is colourless and tasteless and can form an air-tight seal. Today a multi-million pound industry revolves around the production of cork for bottles — since it is still the only material which through its natural elasticity, can accomodate any faults in the necks of the bottles used.

Over half the world's cork is currently produced in Portugal where most large tracts of forest are state-owned. Despite the trebling of production over the last fifty years and substantial production by Spain, the ever-increasing demand for cork may well out-strip supplies in the future.

Plants on the wall

Plant ingredients are again vital to both wallpaper and household paint.

PAINT. Plant oils such as linseed, tung and soyabean have traditionally formed the basis of many cans of paint, together with mineral pigments, thickeners such as China clay or talc, and other ingredients including resin from various pine trees. Still used widely both for paint and varnish manufacture, the oils are boiled with compounds containing heavy metals such as magnesium, cobalt or lead, which helps them to absorb oxygen (and so dry quickly) after they have been painted onto a surface, and to form a hard film.

Thinning agents such as turpentine, extracted from pine oleoresin and processed pine wood have also been used extensively for the making of paints and as cleaning substances for paint brushes. However, the increased use of cheaper petroleum-based solvents ('white spirits') and the growth of paints based on polymeric latexes has brought about a decline in the use of turpentine. Modern water-soluble latex paints, which are now accounting for an increasing proportion of commercial paint production involve plants in a different way. They are mostly made from polymer resins derived from fatty acids which have themselves been split from vegetable oils. Many paints also contain plant cellulose in soluble form to give them stability and to make application easier. Cellulose gives an even consistency to the mixture, helping the pigment to disperse rapidly. Alginates from seaweeds are similarly employed, improving the paint's flow, suspending pigments and helping good coverage to be achieved. In some vinyl emulsions gum arabic (*Acacia senegal*) controls the consistency of pigments and any settling that occurs.

Wallpaper and paints both contain plant ingredients

WALLPAPER. Most wallpaper, like ordinary writing paper, is made from a base of woodpulp. Some companies are using increasing quantities of recycled paper as a base for their wallpapers, but trees will have been the starting point in most cases. A vast range of woods have been used over the years, with different sources of supply for different countries and the companies within them. The price of pulp from different regions will often determine which tree species are ultimately involved, and as this can vary greatly it is often very difficult to say precisely what each roll of wallpaper is made from. In Britain, softwoods such as spruce and pine species have a long history of use, as well as hardwoods such as birch, beech and aspen, often grown in Northern European countries. North American forests producing a mixture of soft and hardwood trees have also been a traditional source of pulp though today's supplies could be made from eucalyptus trees grown in Brazil or South Africa or other fast growing plantation species.

To make the pulp, the tough cellulose fibres in the wood must be separated and broken down and the lignin which binds them removed. The two ways of doing this in the trade are by chemical and mechanical means. The chemical method, which is more expensive, uses acids or alkalis to dissolve out the lignin (about 20 to 30 per cent of the wood's volume). Because this does not damage the fibres it makes what is referred to as a 'strong' pulp, useful for good quality paper. The mechanical method involves grinding up the wood chips by machine. The heat generated softens the lignin and the fibres are literally torn apart. This process damages the fibres considerably, however, and so the resulting pulp and paper will be much weaker. Most wallpaper on sale in Britain is made from about 60 to 70 per cent mechanical pulp — which provides bulk — and the rest chemical pulp, to give it the strength needed to survive the manufacturing process and to be pasted onto a wall without tearing or dissolving.

The wallpaper paste or glue which sticks each sheet to the wall comprises many plant-derived ingredients too. Most of these are thickeners and binding materials and include potato and maize starch, agar from seaweeds and powdered plant cellulose.

Household accessories from plants

The average house is full of articles made from plant materials; examples include everything from wooden bread boards, cork place mats and rattan dog baskets to household brooms.

HOUSEHOLD BROOMS. A number of brooms and brushes on sale in hardware shops and stores today are made directly from plants. In the past, before brush-making became an industry, locally available materials such as twigs and small branches, grasses and reeds or shrubs like heather and broom were used in different areas for sweeping floors, but the main source of broom fibre was hog bristle.

In the mid-nineteenth century, piassava fibre from Brazil and Mexican fibre, as its name suggests, from Mexico, caught the eye of manufacturers. Piassava fibre was at that time being used for packing round sugar cases and other goods imported into Britain, whilst Mexican fibre was an important stuffing material for sailor's mattresses. They were soon adopted by broom and brush manufacturers and paved the way for the introduction of several other plant fibres from the tropics. Cocos fibre from coconut husks and kitool and palmyra fibres from other palms, gave manufacturers a large choice of new materials.

Today, despite the impact of synthetic substitutes, plant fibres — especially those from palms — remain an important raw material for brush and broom making. Around the world 10 000 tonnes of fibre are harvested from palms alone each year for this purpose, earning appreciable amounts of foreign exchange for some developing countries.

Fibres currently used for brooms and brushes in Britain include Para piassava from the palm *Leopoldinia piassaba* and kitool fibre from the Jaggery palm, *Caryota urens*. The dried flowering stems of a sorghum species (*Sorghum bicolor*) meanwhile, provide the raw material known as broom corn or Italian whisk.

Good broom fibres, for example Italian whisk, should have plenty of natural spring in them and, especially for heavy use, should be able to recover their normal position immediately after sweeping. Palm fibres are particularly well suited to the job. They are extracted from the portion of the palm leaf stalk that remains on the tree after the leaves have been removed or have fallen and are extremely resilient and strong. Many will withstand constant use and immersion in hot or corrosive liquids. They make excellent scrubbing and nail brushes as well as yard or outside brooms and are generally long-lasting.

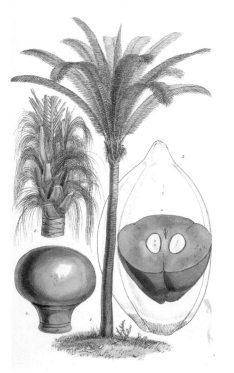

Piassava fibres, from the leaf stalks of the palm Attalea funifera *are amongst the most important of those still used for brush and broom making. This species is native to coastal Brazil.*

HOME AND ABROAD — TRADITIONAL HOUSES

Traditional architecture almost everywhere in the world relies on plant materials to some degree or other. Some of the simplest huts and shelters, most extensive family houses and elaborate ceremonial buildings have all used plants so successfully to fulfil design requirements and to suit the needs of people using them that their form has stayed virtually unchanged for centuries.

Timbers, branches, stems, leaves and fibres skilfully and ingeniously assembled in varying combinations from an enormous range of plants, house and protect the vast proportion of the world's population. In areas of tropical forest especially, the range of building materials used by different peoples reflects the abundance and diversity of plant life around them — but two forest plants in particular stand out: bamboo (tropical grasses) and palm tree species. Between them they have provided an extraordinary variety of timbers and fibres, poles and thatch to suit almost every need.

Wherever people's houses have been made, whether in the foothills of the Himalayas or the Australian 'bush', those materials most readily available and most practical for the purpose in hand have been utilized.

BAMBOO

To date 840 species of bamboo, members of the grass family, have been identified. They are native to tropical, subtropical and some warm temperate regions of the world, but are especially diverse in South and South East Asia. Some bamboos grow taller than many trees and have stems that are stronger than a number of timbers. They grow exceptionally fast (some increasing in size by almost 4 centimetres an hour) and many species reach their full height in only 6 to 8 weeks. In the wild the larger species form dense often impenetrable forests, while the smaller species, equally dense, remain as undergrowth.

Bamboo is used by more than half the world's population every day. The plants can serve almost the entire needs of a community — over 1000 different products have been made from bamboo stems and leaves. The stronger stems (or culms) are used for house construction, scaffolding and bridge-building as well as for water pipes and storage vessels, the canes for furniture, the shoots of many species are eaten as a vegetable and the leaves used for thatch. Fibres peeled from the split culms are used to make rope and twine (useful for fastening panels and poles into place), and these fibres are three times stronger than hemp.

As a building material bamboo plays an immensely important role in almost every country in which it flourishes. In Burma and Bangladesh today, half the houses are made almost entirely from bamboo and in Java woven bamboo mats and screens commonly fill timber house frames. Elsewhere in Asia many rural and forest dwelling

Bamboo scaffolding

peoples make use of the plant in one form or another — often for the supporting structures of their homes, for flooring made from the split and flattened culms, or for making separating walls and screens. In Japan square culms produced by placing a wooden mould over the shoots as they develop are often used as decorative alcove posts for the traditional recess, the tokonoma, built into Japanese houses.

The utility of bamboo culms extends further to the manufacture of high quality veneers. Laminated furniture made from bamboo which has been peeled and processed, then stuck together to make plywood sheets or onto other boards, is now very popular in China and Japan.

The Marsh Arabs of Iraq build their spectacular houses from reeds

Reeds for houses

THE MARSH ARABS. The Marsh Arabs of Iraq have for centuries built their houses — entire villages in fact — from the huge reeds that dominate the landscape they occupy. Sturdy and often very beautiful, the framework, walls, rooves and flooring of these houses are made entirely from reeds, chiefly *Phragmites australis* and *Arundo donax*. The reeds, which cover large tracts of marsh and shallow swampland in Southern Iraq are perennial plants with tall flat, tapering leaf blades. Each year large areas are burnt to encourage new growth, and the young reeds which spring up provide pasture for the Arabs' water buffaloes and other livestock. The mature reeds are used for house construction and the provision of mats and screens.

The tall stems, which may grow to 3 metres in height, are bound together in bundles after cutting and bent into arched pillars to provide the framework of houses, storage buildings and animal pens, whilst walls and roofs are made from reed matting.

Reeds, it seems, may have been the material used to make the earliest pointed arches in the world. In Iraq and elsewhere in the Middle East where wood is scarce, reeds are used as supports for various mud constructions, and notably for arches. Long after the reeds have rotted away, the arch, made of layers of baked mud, still stands. One of the oldest arches in the history of architecture at Gesiphon near the site of the ancient city of Babylon is still standing, its baked bricks more than 2000 years old. The shape of this arch, midway between the rounded Norman and the pointed Gothic arch has the same shape as the entrance to a typical Iraqi diwan (reception room) that forms a part of the Marsh Arab's home. Reeds could well have been the first supports for this unusual arch.

Baka Pygmy building a mongulu (dome-shaped hut) from a frame of saplings covered with giant leaves

Rainforest homes

BAKA PYGMIES. The Baka Pygmies of West Africa construct their dome-shaped huts (mongulu) and short-term, three-sided shelters (lobembe) from an interwoven frame of springy saplings. Various species are used including *Diospyros canaliculata* (mboloa) and these are covered with the giant leaves of *Megaphrynium macrostachyum* (mongongo), the stalks of which are hooked onto the framework by bending them double. The leaves are lapped over one another like shingles, working from the bottom upwards thus making the hut waterproof. Piles of branches are then usually placed on top to hold the leaves down in winds or heavy downpours. The doorway of the hut usually faces those of friends or close relatives but may be repositioned as relationships between the members of the group change.

After a month or two, when local game has been depleted and other forest foods such as nuts, fruit and honey have become scarce, the Pygmy camp moves on. Important social events such as births, deaths, marriages or the start of puberty will also prompt a change of location.

Like other rainforest peoples, most of the implements and possessions used by the Pygmies from spear handles to musical instruments are made from different woods or materials from forest plants. The Baka generally make their beds, for example, from the bark of two trees, *Triplochiton scleroxylon* and *Enantia chlorantha*. After it has been prized from the tree the bark is pounded flat and dries into a hard rigid board. This is then raised off the ground on stakes and covered with leafy bedding comprising ferns or herbaceous plants. To illuminate their evenings the Baka make resin torches, slashing the bark of several tree species including *Canarium schweinfurthii*, leaving them to bleed, and then moulding the soft fresh resin into shape before it sets hard.

Quichua Indian house in coastal Ecuador viewed across jungle garden

COMMUNAL HOUSES. The small single family houses used by many rainforest peoples are a complete contrast to the huge communal houses of various tribal groups in the Amazon rainforest and in Indonesia.

In the north-west Amazon region the Tikuna, Barasana, Witoto and Yukuna Indians have all lived traditionally in large communal malocas. These can be immense, holding up to 150 people and serving a number of functions that go beyond our own conception of a family house. The Yanomami Indians, also from the north-western Amazon, build both malocas and shabonos of various types. Also very large, shabonos take the form of rounded or conical shelters which are either entire or comprise separate screens that form a circle; alternatively they may be made of large gabled huts also positioned to make a circle. The building materials, gathered and erected by the men, consist of strong palm and other timbers, vines for tying, palm leaf thatch for roofing and often flattened sheets of bark for making walls.

Each Indian group has its own style of architecture as one mark of its identity. Within this general style even small variations in building technique (such as the way in which roofing leaves are woven into thatch) will often distinguish subdivisions in the group. Many of these differences in building styles, however, are variations on a common theme. Whilst the malocas of the Yukuna are round, those of the Witoto oval and the Tikuna's are rectangular, all are built around four central pillars which support a steep thatched roof that is tent shaped.

Several thousand kilometres to the south-east the peoples of the Upper Xingu river live in vast houses whose roofs and walls curve continuously like an aircraft hangar. Majestic, very beautiful constructions, a typical Mehinaku house will be some 30 to 40 metres long, 16 metres wide and 9 metres high. Its framework which is made of very flexible wood called pindaíba (taken from various trees belonging to the Annonaceae family) forms a graceful oval dome, which is thatched with bunches of dried satintail grass (*Imperata brasiliensis*). Each house is an architectural masterpiece, with a perfect ventilation system created by the overlapping roof; the interior is cool in summer yet warm during the cold winter nights. As many as thirty people will live in each of these Mehinaku buildings, which are generally separated from each other by distances of several kilometres. For each person the communal house is the centre of his or her world. It serves as a family home and kitchen (each family will have its own private space and fire), dining room, meeting place, workshop and storehouse, dance hall and spiritual focus. The house and each feature of its design represent in fact both the structure and divisions of their own society and a model of the cosmos and spirit world which links them intimately to their outer forest home.

Inside the Mehinaku house hammocks used for sleeping are made from fibres extracted from the leaves of the buriti palm (*Mauritia flexuosa*). The silky fibres make a tough cord used not only for weaving hammocks but for many other household purposes. Only the most tender buriti leaves are chosen by the women for hammock making. They are gathered by canoe from the swamp lands and river banks on which they flourish and shredded to make the cord. A strong but soft and comfortable hammock is a prized possession amongst the Mehinaku. Men will often call their wives 'my little hammock' as a term of endearment!

In Kalimantan, the Indonesian part of Borneo, Dyak peoples (a name which refers to the many different groups of people who have lived on the island since the earliest times), have traditionally lived in long houses. Described as the 'ultimate in communal living', these houses can accommodate up to one hundred families. Each has its own front door and private area inside which comprises sleeping accommodation and storeroom, and at the back of the house a kitchen. The front of the building has a wide, open, porch-like area running its length where social and communal activities take place. The long houses are often built almost entirely of bamboo, using the stems for structural support, walls and flooring.

Sadly, successive unsympathetic administrations in Kalimantan have worked to break up these communal houses, regarding them as a barrier to 'progress'. As Robin Hanbury-Tenison has explained 'there is the puritanical belief that it is somehow immoral for families to

Mehinaku Indian house from the Xingu region thatched with satintail grass. These buildings can measure 30 to 40 metres long, 16 metres wide and 9 metres high

Long house in Borneo built by Dyak peoples. Such buildings can accommodate up to 100 families

live in such close association and that all sorts of licentious behaviour and disgusting orgies will take place if people are not separated by walls and space at night. It is assumed that the only normal, healthy way of life is the single family unit; that hygiene, social responsibility and work will suffer under communal living conditions'.*

Not so far away on the island of Sulawesi the Toraja people have also developed a spectacular and beautiful style of architecture. Tall, narrow, family houses with extraordinary curved roofs stand on sturdy wooden piles cut from tree trunks. Carved and painted wall panels and doors are made of planking cut traditionally from uru wood which is both rot-resistant and finely grained, whilst the roof is composed of split bamboo tiles which are threaded together using strips of palm fibre and tied in place. A great number of layers (around 60 in some houses), the upper overhanging the lower in succession form this very dramatic and distinctive structure. The upper curve of the roof is shaped to resemble the outline of a boat, in reference to the traditional importance of this means of transport. The Toraja believe that their ancestors came from southern China or Cambodia by boat and their houses are also sited facing north to mark the event.

Supernatural powers are attributed to ancient houses, which form the hub of the Toraja's social ordering and are the fixed point from which people calculate their relationships with others.

RICE BARN

The rich carving and ornamentation of the Toraja rice barn — made entirely from forest materials — displays the wealth of a particular family and the number of rice fields it owns. Used to store the rice crop and seed for next year's planting, the barns are central to the social and cultural orientation of the Toraja and strict rules govern their use. Bamboo culms split into sections and overlapped to form an impressive water-tight roof are a distinctive feature of both barns and family houses.

Toraja rice barn

---

* R. Hanbury-Tenison, *Worlds Apart*, Granada, London 1984.

Man picking medicinal plants in Ecuadorian herb garden

Your very good health — plants that cure us

When we reach inside the medicine box we are often reaching indirectly for a plant! For the cut toe, throbbing head or cough and cold as well as many of the more serious complaints that need hospital treatment, plants have given us effective modern cures.

Our widespread use of packaged, brand-named medicines to help us combat everything from hay fever to heart disease has seldom led us to believe that plants could be involved. In modern Western medicine, plants have tended to be seen as quaint relics of the past and next to useless when it comes to treating serious illness. However, one in four of all the prescription drugs dispensed by our local chemists is likely to contain ingredients derived from plants, and the commercial value of these products is estimated at around $40 billion every year!

Plants have in fact given our modern Western pharmacopoeia some 7000 different medical compounds. The array of medicines derived from them is impressive — heart drugs, analgesics, anaesthetics, antibiotics, anti-cancer and anti-parasitic compounds, anti-inflammatory drugs, oral contraceptives, hormones, ulcer treatments, laxatives and diuretics — but a surprisingly small number of plants is involved. The 120 or so plant-based drugs prescribed for use world-wide come from just 95 species. When we look at the statistics of medical research involving them the reasons begin to emerge. Of the 250 000 species of flowering plants, only some 5000 have had their pharmaceutical potential tested in laboratories and very few have been acknowledged in the West to have any real therapeutic value. Added to this is the $50 to $100 million needed to make a new drug and the ten years it takes to develop it. Western medicine, however, treats only a tiny proportion of all the people on earth. Eighty per cent of the world's population relies entirely on local medicines made almost exclusively from plants. It is estimated that between 35 000 and 70 000 different species have been used as medicines by various peoples of the world and as Western drugs continue to be unaffordable or in many instances inappropriate, this pattern of dependence on local herbal cures is set to rise.

In recognition of this fact and since the two should ideally complement each other, the World Health Organization (WHO) has been attempting to incorporate traditional medicine officially into the health care systems of developing countries. In 1988 the International Union for the Conservation of Nature (IUCN) and the World Wide Fund for Nature (WWF), together with WHO brought health professionals and leading conservationists together in Thailand to produce a set of guidelines for countries wishing to make the best use of their medicinal plants and conserve them for the future. They stressed the vital importance of these plants in primary health care and the great potential of the plant kingdom to provide new drugs.

Several countries already boast systems in which traditional plant-based medicine is officially accepted. In China, which has several botanic gardens devoted entirely to medicinal plants, some 5000 plants are used regularly by 800 million people. The Russians, similarly, never developed the contempt for folk medicine that exists in the West and continue to rely on plants for most of their cures. In India and Sri Lanka, Ayurvedic medicine, which treats disease as a state of disharmony in the body and addresses physical and spiritual ill health

Parsley Pert.

Cow Parsnip.

Ground Pine.

Wild Poppy.

Illustrations from Culpepper's Herbal, *first printed in 1652*

A SHORT 'HISTORIE' OF MEDICINAL PLANTS

For most of our history plants have been the principal source of the drugs used to cure and prevent illness. All around the world, people belonging to the most advanced civilizations and the simplest cultures have relied on plants to keep them healthy. Only recently, with the advances in synthetic chemistry have 'developed' countries broken their dependence on cures that came almost entirely from plants. Despite the learning of such classical authorities as Dioscorides, whose 'De Materia Medica', written in the first century AD became the basis of European medicine for the next 1400 years (expounding the virtues of some 600 different plants), the Dark Ages produced a long period of ignorance in Europe from which modern medicine only slowly emerged. People had no idea how plant cures worked and such entrenched and ancient beliefs as the Doctrine of Signatures which held that like cured like (for example, a heart shaped plant could cure a heart affliction) were difficult to change.

The development of the Physic Garden attached to University Faculties of Medicine in the sixteenth and seventeenth centuries and used for teaching purposes as well as the supply of plant-based drugs, played an important role in the interlinked disciplines of botany, pharmacy and medicine, but it was not until the nineteenth century that the curative properties of plant chemicals and their modes of action started to be really understood.

Today the great complexity of many plant compounds essential to modern medicine has made their synthesis impossible or impractical and the plants themselves are our only source. But still, in some areas of Western medicine a rather contemptuous attitude exists towards them.

As the chemist R.S. De Ropp pointed out in 1954 with reference to the chemists and pharmacologists who shared this attitude, they made the error 'of supposing that because they had learned the trick of synthesizing certain substances, they were better chemists than Mother Nature who, besides creating compounds too numerous to mention, also synthesized the aforesaid chemists and pharmacologists'!

together, is dispensed by thousands of practitioners. In Pakistan Unani Tibb, which does the same, is also used alongside Western diagnosis. Traditional healers in many African countries are legally recognized by their governments today and encouraged to work in co-operation with Western medical and scientific personnel. Ordinary people too, especially in rural areas, have inherited generations of medicinal plant lore which often enables them to cure themselves. In countries as distinct as Ecuador and Vietnam many plant their own medicinal herbs and are encouraged to use them by health care officials.

A significant difference exists, however, between the way in which Western science views an illness and its cure and the way it is perceived elsewhere. Many societies believe today, just as they have for centuries, in the power of a spirit world to influence them and their environment. As such, the distinction between an illness that is purely 'physical' and one that has a spiritual cause is often unclear to outsiders and the two are difficult to separate. Ailments of many kinds are thought by some to arise from an imbalance in the body's equilibrium, which may in turn be caused by supernatural intervention or the breaking of an old taboo. Whilst plants are widely used as the apparent cure, the patient's recovery often seems to be determined both by his own and his healer's psychological powers, in combination with the

Man from the Andes with bunch of medicinal herbs

PLANTS AND MAGIC

In many countries of the world where traditional medicine is practised, illness is believed to be able to attack the body only when in a specific state of weakness. Improper care of the body, worry, fear, temperamental or environmental difficulties could all cause illness, but the breaking of some religious taboo or an affliction caused by an angry ancestor or other spirits is often held responsible. Often the patient is said to suffer from soul loss caused by one or more of the above.

To establish the true nature of the complaint the medical practitioner listens carefully to the patient and examines him, but divination is often essential. In parts of Peru, examination of the entrails of a guinea pig or chicken, or the yolk of an egg which has been passed over the body of the patient is carried out to elucidate the problem. In many African countries, a set of numbered seeds, shells, animal bones or carved pieces of wood are thrown to the ground and the pattern they form is 'read' to indicate both the problem and the necessary treatment.

In general the patient is seen as a whole and the cure involves restoring some imbalance which has allowed the illness to take hold. Manipulation of the spirit world alongside the use of plant-derived medicines is very widely practised and many different plants are often involved. As medicines they may be taken in the form of infusions (in hot or cold water), applied directly to the skin, or administered in the form of ointment or powders. In the highlands of Ecuador

traditional healers, curanderas, brush bunches of medicinal plants across the body of the person suffering from some spiritually induced complaint, transferring the harm from the patient and giving some protection for the future. The smoke from burning incense and special 'magical plants' has the same effect and libations as well as the consumption of herbal teas are all part of the cure. The Ombiasu in Madagascar similarly oversees the curing of his patients by ensuring that a suitable offering is made to the ancestor spirits who may have caused an illness, whilst preparing potent plant-based medicines.

Though the 'magical' side of these cures has led us to doubt their effect at all, the people who dispense them often have immense botanical knowledge and are experts in their field. As Professor Shultes has written '.... the shaman or witch doctor, usually an accomplished botanist, represents probably the oldest professional man in the evolution of human culture'. Our inability to accept the work of traditional healers has prejudiced our recognition of the validity of thousands of plants as medicines. It is interesting to reflect that the working of some of our own Western drugs is only partially understood. Similarly it has often been found that whilst the effect of a plant may be marked when all its compounds are used together, isolation of those thought to be responsible does not always produce a cure. Whilst it is important that the use and dosage of medicinal plants is examined and refined as we incorporate them into modern treatments, a study of the way in which they have been traditionally used must play a larger role in future medicine.

Datura (Brugmansia sp.) a powerful 'magic' herb used in Ecuador

special attributes of the plants themselves. In such cases, where ritual is heavily involved and the mind appears to be the arbiter of sickness or good health, with plants acting almost as mnemonics, Western science is often at a loss to explain the results.

In other cases, where plant-derived medicines are used directly to cure specific 'physical' ailments or produce a measurable effect, scientific explanations of how they work, based on a knowledge of their chemical constituents are easier. For much of modern Western medicine relies on chemical analysis. All plants are by nature living chemical factories, and it is the complex substances that they produce that we have mostly used to help keep us healthy.

Plant chemicals are employed in three main ways in Western medicine today: the first is by incorporating them directly into medicines and medications, the second uses chemical compounds as blueprints or starting points for the manufacture of new or synthetic drugs and medicines, and the third uses plant chemicals as tools to help us understand physiological and pharmacological mechanisms, especially in drug development and testing.

Interior of 'Ekpe' religious house in Cameroon with leaves used in magic healing ceremonies

PLANTS IN THE FAMILY MEDICINE CHEST

Sphagnum moss with its absorbent and antibiotic activity was once used to treat wounds

Healing a cut with plants

If you cut your finger recently it is likely that the cotton bush came quickly to the rescue along with several different species of pine tree, a variety of soft and hardwood logs and a number of seaweeds!

The first treatment we might use is cotton wool, made as its name suggests, from cotton. The fluffy white mass is made up of the long silky hairs that cover ripe cotton seeds, intended by nature to disperse them.

The natural absorbency of cotton wool, sterilized for medicinal use, makes it perfect for drying up the flow of blood from minor wounds. Though enormous quantities are used, the current high price of cotton has promoted the development of cotton wool substitutes, for use both in the home and as surgeons' swabs. Plants are still involved, as these substitutes are generally made from viscose, itself derived from a large selection of hard and softwood trees, including pine and eucalyptus species, from around the world.

A plant no longer used in modern medicine has been of major importance in the past for dressing wounds. Sphagnum moss (*Sphagnum* species), able to absorb sixteen times its own weight in liquids, and more than twice as much moisture as cotton, helped staunch the bleeding of injured soldiers during both World Wars. Long famous for its antibiotic activity and wound-healing capacity, analyses of the moss have revealed the presence of several associated micro-organisms, including *Penicillium*, which seems to be responsible. Our Bronze Age ancestors in Britain made use of the extraordinary capacity of sphagnum moss for healing injuries, as did the Lapps, some North American Indians and Inuit groups until very recently.

Surgical dressing derived from seaweed

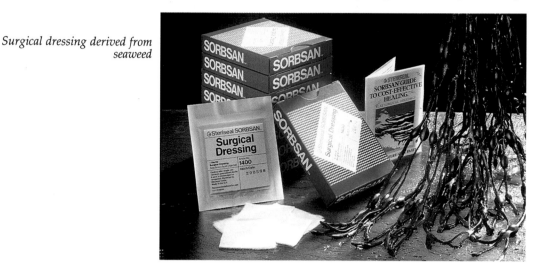

Bandages

For bandages of all descriptions, the cotton plant again has given us a prime material. Cotton has also been used to make the wound dressing known as lint, though as this name (an abbreviation of the Latin term *linteus* 'of linen') indicates, the flax plant has been the traditional source. By definition lint is linen cloth scraped so that its surface is soft and fluffy or a soft fuzzy substance made up of short yarn or fabric fibres.

Though not so likely to be used at home, some special dressings have been devised to help clean and heal heavily soiled or very severe wounds. Amongst these is activated charcoal cloth which is used to help clean dirty, heavy cuts and lesions and to clear up infection. The base material, viscose, is charred at very high temperature to form charcoal, which effectively absorbs bacterial toxins from the wound and the odour they produce. Calcium alginate dressings meanwhile which have a haemostatic effect can be left in the body in cases of severe injury. Gauze-like in appearance, the dressings are made from alginates derived from seaweeds. On contact with the wound, they dissolve slowly and are absorbed into the body as it heals. The seaweeds used for the commercial production of alginates belong to the Phaeophyta or brown algae. Large quantities are gathered from the Atlantic coasts of Europe and North America, and from Japan.

Antiseptics

Many dressings, as well as being sterilized before use are treated with some kind of disinfectant or antiseptic preparation. Though its most common source in Europe is the mineral saltpetre mined in Chile, tincture of iodine or potassium iodine solution is also made from seaweed ash. In Japan the island of Hokkaido has been the main centre of production, and a chief producer of the world's supply. Some two million tonnes of wet seaweed are needed to make 115 tonnes of iodine. The species used will depend on the area of harvest, but the kelps, large brown seaweeds of the order Laminariales have been widely utilized.

WYCH HAZEL

Hamamelis virginiana

With its antiseptic, astringent and haemostatic qualities, wych hazel was used by many North American Indian groups to treat a great many complaints — from back ache to painful tumours and ulcers. The Cherokee Indians made a tea from the leaves and bark which they drank to alleviate colds, fevers and sore throats and also used to wash sores and wounds. To treat scratches they bruised the leaves and rubbed them onto the skin. Numerous varieties of wych hazel are now commonly grown in gardens as ornamental trees.

Another common antiseptic, used also in treatments for troubled skins, is wych hazel (*Hamamelis virginiana*). When distilled with alcohol the leaves and bark of *H. virginiana* yield an extract which helps prevent inflammation and controls bleeding. Over one million gallons of wych hazel are sold in the United States alone each year.

The distinctive pine smell of some familiar household antiseptics derives mainly from the inclusion of terpineol or pine oil, which serve as solvents and coupling agents for other ingredients and which contribute valuable antiseptic and bactericidal properties in their own right. Animals too, have benefitted from the pine tree. Preparations made from pine tar are available to protect the hooves of farm animals from foot rot.

Fabric-backed sticking plasters are made almost entirely from plants

Sticking plasters

Most fabric-backed sticking plasters are an interesting amalgamation of materials made almost entirely from plants. Though the soft padding or filler material is likely to be made of polyester (derived from oil) and the fine netting that keeps the padding clear of the wound from plastic, the fabric is usually made of cotton, or rayon (made from woodpulp) mixed with cotton. Plants also keep the plaster where you put it. The traditional base of the adhesive which coats the underside of the plaster and provides elasticity is latex from the rubber tree *Hevea brasiliensis*. Thousands of tonnes of natural rubber are used each year by the world's manufacturers of sticking plasters for this sole purpose.

Although a dried film of natural rubber is auto-adhesive — that is it will stick to itself — it is quite dry to the touch and will not stick to other materials. When mixed with rosin and certain rosin derivatives, and other 'tackifying' resins, natural rubber becomes very sticky and will wet and adhere strongly to all kinds of surfaces. Natural rubber and rosin together provide the adhesive mass for the medicated sticking plasters to be found in practically every medicine cupboard.

The gum rosin often used comes from the oleo-resin taken from living pine trees in much the same way as latex is tapped from rubber trees. The oleo-resin comprises rosin and turpentine, and the rosin is recovered by distilling off the volatile turpentine.

Much of the gum rosin used today comes from China and Portugal, but several South American countries have rapidly developing industries. Other important sources of rosin are decayed pine wood stumps which, in America, are solvent-extracted to produce wood rosin, and the sulphate (kraft) pine wood pulping process, from which tall oil rosin is recovered as a valuable by-product.

Resin being tapped from a pine tree

Rosin and its chemical derivatives and the polyterpine resins from turpentine also make a major contribution as tackifiers to many different types of adhesive composition; for example they help to hold the pressure-sensitive sticky price labels onto most of the packaged goods we buy, and the printed paper labels onto tinned foods.

Coughs and colds

During 1987–88 British people spent £86 million on medicines to help alleviate the symptoms of their colds and coughs.

As decongestants and soothing counter-irritants, a large number of plant extracts are available in syrup, pill and powder form today. Lemon, eucalyptus, and cinnamon leaf oils as

---

EUCALYPTUS

Eucalyptus spp.

Eucalyptus oil which has powerful antiseptic properties is distilled from the fresh leaves of a number of eucalyptus species, particularly *Eucalyptus globulus*, grown specially for the purpose in their native Australia, as well as Brazil, the Congo, and parts of the Mediterranean.

The main constituent of the oil, accounting for some 70 per cent of its volume, is eucalyptol, also known as cineol. It is much used as a counter irritant in throat pastilles and cough syrups, as an antiseptic in gargles, ointments and linaments and as an ingredient of inhalants used to relieve bronchitis and asthma.

Other oils, not used medicinally, are also produced by eucalyptus trees. Whilst some have found industrial applications, the aromatic eucalyptus oils are important in the perfumery trade. Oil from the leaves of *Eucalyptus citriodora* emit a delightful lemon scent, whilst that from *E. sturtiana* is reminiscent of ripe apples.

Eucalyptus globulus

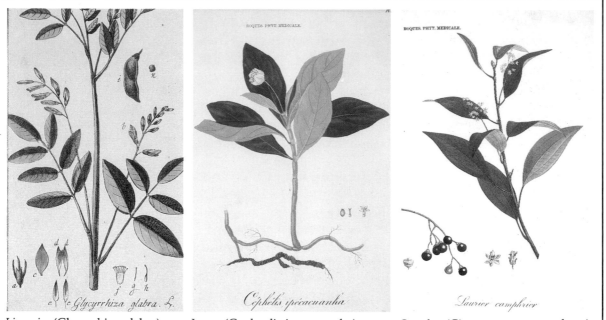

Liquorice (Glycyrrhiza glabra)   Ipecac (Cephaelis ipecacuanha)   Camphor (Cinnamomum camphora)

LIQUORICE

Glycyrrhiza spp.

Liquorice is obtained from the dried roots and rhizomes of several *Glycyrrhiza* species, in particular *G. glabra* — all perennial herbaceous shrubs indigenous to southern Europe and Asia Minor. Most of the world's supply comes from plants gathered from the wild in the USSR, Spain, Turkey, Syria, Iraq and Afghanistan, though they were once extensively grown in Yorkshire. Around 50 000 tonnes of liquorice extract are produced each year.

The main active ingredients in the plants are saponin-like glycosides, of which glycyrrhizin (between 5 and 10 per cent) is the most important. In crude extract form liquorice is used as an expectorant and anti-inflammatory, and is common in cough syrups, sweets and pastilles. Liquorice is also used to disguise the bitter taste of other medicines. Many non-medicinal foods and drinks are flavoured with liquorice, as well as tobacco, toothpaste and breath fresheners. It is also often present as a foam stabilizer in fire extinguishers!

IPECAC

Cephaelis ipecacuanha

About 100 tonnes of ipecac are produced each year from *Cephaelis ipecacuanha* — mainly grown in Nicaragua, Brazil and India — for treating coughs, bronchitis and amoebic dysentery, and for use as a powerful emetic.

The bark of the thickened, annulated roots which should be collected when the plant is in flower contains most of the active ingredients — the alkaloids emetine, cephaline and psychotrine. These same alkaloids cured Louis XIV of dysentery in the seventeenth century and today are valued for the treatment of other parasitic infections such as bilharziasis and guinea worms. The popularity of ipecac in traditional medicine as a cancer cure may also be well founded. The Guaraní of Brazil and Paraguay were the first to discover the virtues of ipecac; their term 'i-pe-kaa-gueñe' (road-side sick-making plant) gave rise to the Portuguese name ipecacuanha.

CAMPHOR

Cinnamomum camphora

The camphor tree, an evergreen native to China, Japan and Taiwan has been used for centuries to help alleviate cold symptoms. Its distilled wood yields camphor and white oil of camphor, both of which are slightly antiseptic and have expectorant and analgesic properties. Most often applied externally — as in camphorated oil — rheumatic pain and inflammation, fibrositis and neuralgia have also been relieved by its use. Some disinfectants incorporate camphor oil in their formulation too.

Whilst most natural camphor oil is currently exported from Taiwan, synthetic camphor derived from coal tar and turpentine is now a common substitute for the plant-derived compound.

well as thymol from thyme and menthol from mint species are all very widely used to relieve colds. The soothing properties of extracts from liquorice roots and lobelia leaves soothe bronchial spasms and sore throats.

A plant from the Amazon rainforest also yields a powerful expectorant for the treatment of bronchitis, coughs and whooping cough. The dried or powdered roots of two species of *Cephaelis*, herbaceous shrubs, one native to Brazil, the other indigenous to Panama, Colombia and Nicaragua produce the drug known as ipecac or ipecacuanha. This contains a number of powerful alkaloids of which emetine and psychotine are most important. Ipecac is useful in its crude extract form as an expectorant, and very important in larger doses as the standard emetic given to children who have taken overdoses of pills or swallowed poisonous substances. Its alkaloids, especially emetine, are also a specific cure for amoebic dysentery. More than 7 million prescriptions dispensed this year in the USA are likely to contain ipecac in some form or other.

Stirring an extract of squill (Urginea sp.), used in cough linctus preparations, in a Victorian copper evaporating dish

Dried roots of Cephaelis *shrubs produce ipecac, used to treat coughs*

The blueprint for aspirins came from the bark of the white willow (Salix alba)

The willow tree to soothe an aching head

Without the willow tree, the aspirin, probably the world's most widely used drug, could not have been developed. Though aspirins are made today from man-made compounds, it was substances extracted from the leaves and bark of the white willow (*Salix alba*), and later from the perennial herb meadowsweet (*Filipendula ulmaria*) which gave chemists the blueprint for this famous headache cure.

The willow has a very long history of use as a pain-killer. Observation of its leaves as they moved in the wind — thought to resemble the trembling of a fever patient — led to the ancient use of infusions of white willow bark as a general treatment for fevers. By the time of Dioscorides (first century AD) the willow was also being used to treat gout, rheumatic pains, toothache, earache and headaches. It was not until 1827, however, that salicin, the active ingredient responsible for the alleviation of pain, was isolated. The fragrant-flowered meadowsweet, meanwhile, common throughout Europe and Eurasia was found to yield a similar compound — salicylic acid. Whilst salicin became the focus of experimentation, acetyl salicylic

Garlic — *Allium sativum*

Allium cepa — the onion

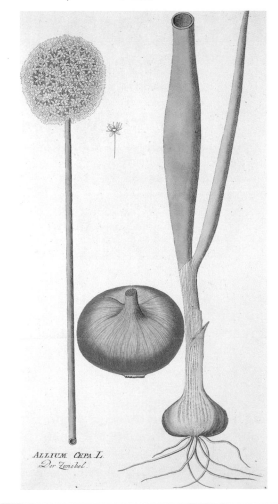

ALLIUM CEPA L
Der Zwiebel.

FLU PREVENTION WITH GARLIC — NATURE'S MOST PUNGENT ANTIBIOTIC!

As a natural antibiotic, garlic, a member of the lily family, has been used in many parts of the world to treat and very often prevent all manner of infections, especially those affecting the nose, ears, chest and throat.

Colds, 'flu, chest congestion, sinus problems, tonsil infections, whooping cough and bronchitis as well as intestinal disorders, high blood pressure, indigestion, acne, asthma and diptheria have all been relieved with the help of garlic. For the early treatment of colds and 'flu two cloves of garlic kept in the mouth, one on each side and frequently renewed, is a much praised folk remedy. Slight crushing of the cloves to release their juice is said to make the treatment more effective.

The active ingredient in garlic which is responsible for its tenacious smell, is the chemical allicin. It effectively suffocates the bacteria causing infection, but unlike more powerful commercial antibiotics, which tend to destroy 'friendly' bacteria as well as the dangerous ones, allicin kills only harmful germs.

Intended by nature to provide food for the growing garlic plant, the bulb has a very long history of use both as a human food and medicine. By at least 3000 BC Mediterranean and Far Eastern peoples had discovered its virtues and the plants (which probably originated in Central Asia) were soon distributed around the world.

The Greek physicians Hippocrates and Galen whose teachings greatly influenced medical science for many centuries, prescribed garlic for intestinal problems and infectious diseases, whilst the Roman scholar Pliny believed the bulb could cure respiratory, bronchial and tubercular conditions. Dioscorides, physician to the Roman Army, also used garlic to treat worms.

Nearly all the members of the *Allium* genus to which garlic and its relatives belong, and in particular the onion, leek and chive have been reported to have medicinal properties. Onions, thought to have originated in Central Asia, are taken for granted by most of us today, but were prized in antiquity by many peoples of the East and notably the Egyptians, who grew them as long ago as 3200 BC. Today their medicinal and nutritional values are being re-examined and their importance as major constituents of some rural diets reappraised. Whilst raw onions are most effective, even a small portion of cooked onion is said to be beneficial in lowering the cholesterol level in the blood. Onions are also said to stimulate insulin production, promote the healing of wounds and suppress allergies.

acid, produced in Germany in 1899 by combining salicylic acid and acetic acid proved to be more effective and gave quick relief for all kinds of pain. The aspirin thus arose, taking its name from the Latin name for meadowsweet in use at the time: *Spiraea ulmaria*.

The white willow (Salix alba) *grows to heights of 18 to 24 metres with a girth of some 6 metres. The ancient Greeks used extracts of willow bark to relieve pain, as did groups of North American Indians until relatively recently*

Opium poppy with cut pod dripping latex

Killing pain with poppies

Another very well known plant, but one that has come to have some much more sinister associations, the opium poppy (*Papaver somniferum*) yields morphine, the greatest natural painkiller of all. Morphine is just one of some 25 alkaloids which are present in raw opium, the milky latex exuded by the immature seed capsule of the poppy when cut. It is also extracted from the stems and leaves of the plant.

Able to relieve pain and resulting sleeplessness, and calm anxiety, morphine takes its name from Morpheus, the Greek god of dreams. The Greeks regarded sleep as the greatest healer and the poppy was celebrated in their art, literature and religion from the earliest times. Morphine was also used as long ago as 6000 BC by Swiss Lake Dwellers and by the Sumerians (4500 BC) in the form of small balls of opium which were eaten or mixed with wine.

Today, the powerful analgesic properties of morphine are used in medicine to relieve severe pain. Cancer sufferers, those with heart or lung failure and people suffering traumatic shock are all likely to be given morphine. In combination with other drugs it is also used to relieve renal and intestinal colic and angina.

Codeine, another of the alkaloids contained in the poppy (only one fifth as strong as morphine for analgesic use), is widely available as a 'household' painkiller, especially as a cough suppressant, in cold remedies and for the relief of all kinds of bodily aches and pains.

Some 200 tonnes of opium alkaloids are used in medicines around the world each year. Since morphine still cannot be synthesized chemically, opium poppies are the only source of its supply. Codeine is both extracted from morphine and isolated directly from harvested poppy leaves and stems. Native to Asia Minor, this attractive perennial with red, pink or white flowers, is a prime example of the dangerous duality of some of nature's most useful medicinal plants. Opium is highly addictive, and as the source of heroin, can be a deadly and destructive drug when misused.

Subsequent to its isolation from crude opium in 1805, British physicians who prescribed morphine in medicines like laudanum (now called tincture of opium) for such famous patients as Samuel Taylor Coleridge and Elizabeth Browning, caused or contributed unwittingly to their addiction to the drug. Apart from its soothing action, a further attraction lay in its effect as a mental stimulant, creating the impression of alertness without any inhibition.

Britain played a distressing role in the addiction of thousands of Chinese people to opium in the early nineteenth century. The British East India Company used opium, produced from poppies grown in Burma and India and smuggled into China, to obtain silver bullion with which to pay the Chinese for their much coveted tea and silk; a situation which finally led to the Opium Wars between 1839 and 1842.

Heroin, discovered in 1898 as a result of chemical changes made to the morphine molecule, was found to be more effective than morphine as a painkiller and more effective as a cough suppressant than codeine. It was openly sold in cough suppressants in North America until 1917. It was not until World War II, however, that morphine and heroin addiction became serious social problems in the USA and Europe.

For medicinal use the most important opium producing countries today include India, Turkey, Bulgaria, Yugoslavia, USSR, Australia, France and Spain. Major sources of opium and heroin for the illegal markets in America and Europe come from the Golden Crescent (Iran, Afghanistan and Pakistan); the Golden Triangle (Burma, Thailand and Laos) and Mexico.

Non-narcotic poppy seeds from the opium poppy are familiar to most of us as condiments sprinkled on baked goods and pastries. They also yield an oil useful in paint, varnish and soap manufacture.

Pills from plants

Swallowing an aspirin or any other sort of pill can be unpleasant at the best of times. If it were not for a number of useful binding substances derived from plants which help to hold all the active ingredients together, it would be even more difficult.

Potato starch, fractionated palm and coconut oils, cellulose from wood pulp, gum arabic from species of *Acacia* trees and agar, mannitol, algin and carrageenan all from seaweeds are commonly used for mixing or binding the ingredients.

Alginates in particular have been very useful as stabilizing agents in many preparations. Since they are not attacked by the stomach's digestive juices, tablets which contain them can reach and dissolve in the intestine if this is required, instead of the stomach. Gum arabic, used in the 1920s for the treatment of low blood pressure caused by haemorrhage or surgical shock, is especially useful as a binder of tablets and as an excipient (a substance mixed with a drug to make it easy to administer) in many types of pill coating. For the preparation of cough medicines it produces a smooth viscous syrup — preventing the crystallization of sugars — and is one of the best emulsifying agents for calamine and kaolin emulsions.

Arthritis

As many as 200 different types of arthritis affect millions of people each year, in Britain alone. Despite continuing research in hospitals and universities all over the country there is still no certain cure for these agonizing diseases. Against such a background more and more arthritis sufferers have been turning to homeopathy and treatments largely unaccepted by the medical profession.

FEVERFEW

Tanacetum parthenium

Feverfew was recorded by John Parkinson in his seventeenth century herbal as 'very effectual for all paines in the head'. Modern research into the effects of the leaves when taken by migraine sufferers has led some members of the medical profession to regard feverfew as a specific treatment for migraine. The name feverfew is said to derive from 'febrifuge' — in reference to its tonic and fever-dispelling properties.

Oil extracted from the evening primrose (a name which may refer to any one of about one hundred species of herbaceous plant in the genus *Oenothera*) was used in recent trials conducted at the University of Glasgow's Centre for Rheumatic Disease on arthritis sufferers. Some two-thirds of the patients involved reported an improvement in their condition after taking the oil over a nine month period. Whilst much more research still needs to be done, the future of evening primrose oil for the alleviation of this and many other ailments looks promising.

Apart from its use as a possible remedy for arthritis, evening primrose oil has also been used to help sufferers of migraine, asthma, eczema, high blood pressure, premenstrual tension, Parkinson's disease and multiple sclerosis as well as hyperactive children. Now officially recognized by the medical profession as a treatment for atopic eczema and given a product licence by the Department of Health, oil of evening primrose looks set to play an important role in future health care in the West.

Sufferers of gout (sudden attacks of arthritis caused by the presence of uric acid crystals in the joints) may have been helped by the delicate white or purple flowered autumn crocus (*Colchicum autumnale*). The toxic alkaloid colchicine extracted from the corms and seeds of this European plant, also known as meadow saffron, is currently being used by medical practitioners to treat acute attacks of gout.

Oil of evening primrose is used to relieve many ailments including premenstrual tension, and has recently been used in trials on arthritis sufferers

Baka pygmies use the bark of Pachypodanthium *species (molbo) to treat arthritis*

Ephedra shrubs yield alkaloids used for the treatment of asthma and other allergies

Asthma and allergies

The sufferers of various allergies including hayfever and urticaria (nettle rash), as well as asthmatics whose condition often occurs as an allergic reaction, may well have been helped by various species of *Ephedra* shrubs, which are native to China and Japan.

The green twigs of *Ephedra* are processed to yield two important alkaloids: ephedrine and pseudo ephedrine. Ephedrine is a sympathetic nerve stimulant related to adrenalin and has valuable anti-spasmodic properties. It acts promptly in relieving swellings of the mucous membranes and in clearing both bronchial and nasal passages. Ephedrine has thus become a common ingredient of medicines used to treat allergic conditions.

Ma Huang, as the ephedra shrub is known in China has been used in medicine there for the last 5000 years. A 'tea' made from the twigs was used to treat fever and coughs and also to increase blood pressure. The introduction of ephedrine to Western medicine is relatively recent but it is now a very important multi-purpose drug made both from the plants and synthetically. Other uses include prevention of a fall in blood pressure after spinal anaesthesia in surgery and the stimulation of patients in a coma.

HENBANE

Hyoscyamus niger

A source of the alkaloids hyoscyamine and hyoscine, the henbane plant is one of nature's less pleasant creations — its evil smell and lurid mauve and yellow flowers warn of its very deadly powers if used unwisely.

Every part of the plant is poisonous and neither boiling nor drying removes its toxicity, yet henbane has been used medicinally since the remote past. Dioscorides recommended the milder white-flowered *Hyoscyamus albus* for inducing sleep and easing pains, but less philanthropic individuals used the plant to murder people! In keeping with its unpleasant reputation, Culpepper noted its propensity for growing 'in saturnine places' and recorded that 'whole cart loads of it' were to be found 'near the places where they empty the common Jakes' (lavatories). He also described the use of the smoke produced by burning the plant for healing 'swellings, chilblains or kibes in the hands and feet'. Whilst crushed seeds were a favourite remedy for toothache in the Middle Ages, the use of henbane smoke by scurrilous dentists was notorious. They pretended to cause small worms (in fact henbane seeds or small pieces of lute string) to fall out of the patients teeth into a dish whilst holding their mouths over a brazier — and of course, charged them for the privilege!

Henbane seeds and extract in Victorian collecting jars

Travel sickness

Some very poisonous relatives of the potato: henbane, deadly nightshade, thornapple and the Australian corkwood tree have helped many sufferers of travel sickness reach their destinations without feeling ill! The plants contain alkaloids widely used in medicine today, including atropine and hyoscine.

Though atropine can now be synthesized, the starting point of both this drug and of hyoscine is the dried leaves and flowering tops of plants collected from the wild and grown in cultivation. Deadly nightshade (*Atropa belladonna*) and henbane (*Hyoscyamus niger*) are grown for commercial use in the Balkans and North America, whilst the corkwood tree (*Duboisia* spp.) is cultivated in Australia, producing some 700 to 800 tonnes of dry leaves each year.

Apart from their use in travel sickness pills, atropine and hyoscine are active against such prevalent complaints as stomach and duodenal ulcers and are used by ophthalmologists for dilating the pupil of the eye. Greek ladies used diluted juice squeezed from deadly nightshade to make their eyes look more attractive, as the name belladonna reflects.

Hyoscine, is effective in relaxing spasms of the involuntary muscles, and lessens pain and induces sleep; it is also used to treat several forms of mental illness and delirium, as well as goitre, colic and intestinal problems. It serves too, as a pre-anaesthetic medication.

Contraception — plants for small families

Our best known form of modern birth control, the contraceptive pill, taken by an estimated 80 million women every day, comes to us courtesy of a family of climbing forest vines, the yams.

Though the pill is now mostly made synthetically in the West, it was diosgenin, a steroidal sapogenin that occurs naturally in very high levels in some yam species that enabled this revolutionary means of birth control to be produced. The discoveries of R.E. Marker, an American organic chemist who showed in 1942 how male and female hormones could be made from diosgenin (isolated from a Japanese yam *Dioscorea tokoro*) were at first ignored by the large pharmaceutical companies. But after his discovery of several yam species native to Mexico which had tubers big enough to make commercial production viable, these vegetables became the basis of the world's oral contraceptive industry.

YAMS

Dioscorea spp.

Of some 125 yam species that have been examined for their potential as commercial sources of diosgenin for contraceptive production, three species of the 'Mexican yam' native to South and Central America and grown in North America today are amongst the most important. *Dioscorea deltoidea*, native to India and once very common in the Himalayan foothills has been another major source; since over-exploitation has threatened this yam in the wild, it is now being cultivated in other parts of India especially for diosgenin production. In Japan and China *D. nipponica* is an important source of this compound.

Steroidal drugs have been derived from yams (above), soya beans (pods shown right) and calabar beans (far right)

Steroidal sapogenins are the starting materials for one of the most important group of drugs used in the world today. Alongside oral contraceptives their products include cortisone and hydrocortisone (used to treat rheumatoid arthritis, rheumatic fever, and sciatica), sex hormones and anabolic agents. Whilst these, like the modern contraceptive pill, can now be totally synthesized or made from animal by-products, different plant species but especially the yams are still a major source of their raw materials.

Until recently, diosgenin from yams provided more than 80 per cent of the materials used to make steroidal drugs. Whilst the present level is down to around 50 per cent, this still accounts for the use of more than 200 tonnes of diosgenin each year. In China and India wild yams are still being processed to make oral contraceptives and they will continue to be used, at least in developing countries, for many years to come.

During the last ten years the compounds yielded by a number of quite different plants have also come to provide us with steroidal drugs. The soya bean (*Glycine max*) for example, has been widely used for the compound stigmasterol which is contained in its oil. The calabar bean (*Physostigma venenosum*) a native of West African rainforests is another source of stigmasterol as well as other important alkaloids such as physostigmine (also called eserine), used particularly for eye disorders.

Though its chemical nature and effects are yet to be fully understood, oil from the seeds of the cotton plant could also be a source of oral contraceptives — this time for men! The oil contains gossypol, a bitter, yellowish pigment which until the development of refining processes in the early twentieth century, made the oil very unpleasant to eat. In the Chinese province of Jiangsu, however, cottonseed oil was used for cooking during the 1930s when the processing of cotton in the area made it cheap and readily available. The astonishing result was that no children were born there while the oil was being used. The culprit was discovered to be gossypol. It had worked by interfering with the cells responsible for sperm formation, effectively stopping their development. Whilst compounds made from gossypol have been tested more recently in China in clinical trials, the adverse side effects of the chemical and the apparent social and psychological trauma experienced by some of the male volunteers, indicate that further studies and improvements still need to be made.

Around the world many other plants have been utilized by different peoples to prevent or defer conception, and in some cases to abort a pregnancy. About 4000 plants, approximately half of them from tropical rainforests have been recorded by Western scientists (working largely from the information given to them by local people) as containing anti-fertility compounds. Although only a tiny fraction of all tropical forest species have been screened for their chemical constituents to date, around 260 plants from South America alone have been confirmed as having birth control potential. Professor Richard Evans Schultes, Director of Harvard's Botanical Museum, who has spent forty years studying the native uses of plants in the north-west Amazon has amassed an enormous body of evidence attesting to the indisputable efficiency of plant drugs as used by South American Indians. One group of Indians, the Bara-Makó of the Piraparaná river, make a substance they call 'no children medicine' from a plant of the Moraceae family *Pourouma cecropiifolia*. The bark from the roots of this plant is scraped off and rubbed in water to make a drink, which when taken by women is said to be able to produce permanent sterility.

A tree native to the forests of the eastern Amazon, better known for its yield of one of the world's most durable timbers, is another source of contraceptive compounds. The nut of the majestic greenheart tree (*Ocotea rodiaei*) has long been used as an effective and reliable means of birth control by the women of several Amerindian groups in its native Guyana and its properties are now attracting much attention in the West.

An abortifacient is prepared by women in Nigeria from the fruit of *Lagenaria breviflora*, a rainforest member of the gourd family. Scientific research has now confirmed that extracts from the fruit have both anti-implantation and strong abortifacient properties.

Many remedies for infertility also derive from plants. The Baka pygmies use the young leaf shoots of *Palisota schweinfurthii*, a common herb of the forest understorey, which are chopped and boiled to make a special drink. Asama bark, scraped from the tree *Turraeanthus africanus* is also made into an infusion for drinking. Amongst the Baka, these remedies are prepared by experienced older women whose soothing sympathetic manner and reassuring words create the positive frame of mind that is an important facet of their medical prowess.

Baka pygmies drink the boiled leaf shoots of Palisota schweinfurthii, *and an infusion of asama bark* (Turraeanthus africanus) *to treat infertility*

The rubber tree has given us the only modern contraceptive used by men — the sheath or condom. All sheaths are made from a base of natural rubber, mostly tapped from plantation trees grown in Malaysia though a small amount of latex comes from Thailand.

The great utility of rubber, and the characteristic which makes it the common denominator of all elastomers (products that will stretch whilst keeping all or some of their original shape) is its natural elasticity. In its crude form rubber latex has many properties akin to cows milk. After tapping from the trees and during shipment to the countries where it will be processed, it must be treated very carefully to stop it curdling and decomposing. The surrounding air temperatures are minutely controlled and once the latex has arrived at its final destination, samples will be taken to make sure that its condition is good.

To begin the process which turns the latex into rubber, it is centrifuged using equipment similar to that which separates the cream from milk in modern dairies. Next sulphur is added to the latex and also a metal salt, usually zinc oxide. To this an 'accelerator' (a complex mix of organic salts) is added to promote the reaction between the sulphur and the zinc oxide and therefore the development of the rubber structure. Since natural rubber is sensitive to oxygen and ozone as well as sunlight, more chemical salts are mixed into the base as antioxidants, and antiozonants are added, often in the form of man-made waxes such as paraffin wax. The slightly oily feel of some condoms is due in part to the natural migration of these waxes towards the outside of the rubber composition. To counteract this effect and to help put on the condom, maize and potato starch often coat the outside surface.

Laxatives from plants

It is very likely that any laxative you may use owes its distinctive action to a compound extracted from a plant. Most of the 700 or so laxatives available in the USA and the majority of those available elsewhere are based on plant ingredients. In Britain alone £14 million is spent each year on laxatives.

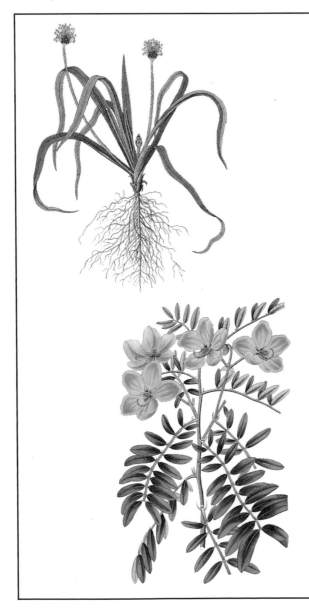

PLANTAIN

Plantago spp.

The ripe seed husks of two common plantains: *Plantago ovata* and *P. afra*, both grown in India on a large commercial scale, are very important to the laxative industry today. The seed or psyllium husks as they are known contain a mucilage which helps cure constipation by attracting and absorbing water once in the digestive tract and in so doing, lubricating it. The mucilage absorbs bacterial toxins too and generally improves peristalsis. India, one of the main suppliers of plantain extracts to companies world-wide produces around 20 000 tonnes each year.

SENNA

Cassia spp.

Millions of kilogrammes of senna are imported each year into the USA, derived for commercial use from the fruit pods and leaves of two of the 500 to 600 *Cassia* species; *C. angustifolia* (known as Indian senna) grown in the Indian state of Tamil Nadu, and *C. senna* (or Alexandrian senna) gathered from the wild in Sudan. *C. senna* is reputed to yield the finest and most valuable variety of senna. It is a small shrub growing to about 0.6 metres high, with long spreading branches. The active constituents, which are used in laxatives around the world are glycosides, also known as sennosides. They increase the peristaltic movement of the colon by their action on the wall of the intestine.

Cassia senna

Traditional age-old remedies for constipation such as rhubarb (chiefly *Rheum palmatum*, not the garden rhubarb), used over 5000 years ago in China, senna from the fruits and leaves of *Cassia senna*, now grown in India and Sudan, castor oil from the castor oil tree (*Ricinus communis*) and extracts from different sorts of aloes are all used in modern laxatives today.

Other plants with notable purgative properties include members of the plantain and buckthorn families. In America bark from the buckthorn tree (*Rhamnus purshiana*) is used to make the laxative known commonly as cascara. Now available in sugar-coated pill form the active compounds in the bark were traditionally taken by North American Indians whilst the properties of the berries (from species native to Britain) were well known to the Anglo-Saxons.

A PINEAPPLE CURE FOR THROMBOSIS

Thrombosis, the blockage of blood vessels by clotting, is responsible for almost half the deaths in developed countries such as Britain each year. Heart attacks, most often caused by a blockage in the blood vessels serving the heart, and strokes which are the result of similar blockages in the brain account for most of these deaths.

The clots that do such fatal damage are made largely of *fibrin*, a protein. Recently, an enzyme (bromelain) found in large quantities in the pineapple has been shown to be capable of breaking down proteins including fibrin and could well play a major part in thrombosis treatment of the future.

Bromelain is currently extracted from the stems of pineapple plants though it is also present in the fruit and leaves. Pineapples were first cultivated by the Guarani Indians of Brazil and Paraguay. Local people drank the juice to aid digestion — especially after eating meat — and as a cure for stomach ache. Pineapple juice was also used to promote the healing of wounds.

PLANTS IN HOSPITAL

Going into hospital for an operation or for whatever treatment the doctor has prescribed is not always easy or painless, but without the help of plants some of our most serious ailments would be much more difficult to cure.

Anaesthetics from the Andes

Surgeons everywhere and of course their patients, have benefitted greatly from the leaves of the coca bush (*Erythroxylum coca*) which have made available some of our most valuable and widely used anaesthetic compounds. The leaves contain some fourteen different alkaloids, including cocaine (present in the highest concentrations) which is currently used as a local anaesthetic in ear, nose and throat surgery. Cocaine has also given us the chemical blueprint for a number of man-made substances used in local anaesthesia including procaine, lignocaine and novocaine.

In its native South America, coca has been grown for centuries by indigenous peoples of the Andes, from Colombia to Bolivia. During Inca times the bush was regarded as sacred and the property of only the ruling Inca caste. Leaves were used for a number of important ceremonial

COCA

Erythroxylum coca

Today, the leaves of the coca bush are chewed by thousands of poor and undernourished Andean peoples to give them stamina. A wad of leaves kept at the side of the mouth and constantly renewed as the leaves lose their potency helps make the carrying of heavy loads and strenuous work of all sorts a good deal more bearable. The compound responsible is cocaine. The illegal manufacture of cocaine as a 'recreational' drug for Western markets has led to the wholesale destruction of many large coca plantations in South American countries, adding to the suffering of those who grow or use the leaves not for profit or to reach a 'high' but to supplement their meagre diets.

purposes, as part of religious observance in the form of offerings and for the divination of omens. The dead began their journey to the afterworld in the company of coca leaves and the plant was revered everywhere. The reason for this special reverence lies with the chemical composition of the leaves. When chewed with a small amount of lime or plant ash, their active compounds are released, stimulating the nervous system in general, enhancing muscle potential, increasing stamina, depressing hunger and generally relieving pain.

Soon after their arrival in Peru and quick to oppress their newly conquered subjects, the Spanish made sure that coca leaves were available to their workforce. Its use allowed them to exploit Indian labour mercilessly, for little food and no remuneration and helped the Indians to endure appalling cruelty and deprivation. Since the Indians then, as now, did not ingest the cocaine alkaloid alone, but in the company of other compounds in the leaves, they did not suffer the addictive or mind-altering side effects associated with the use of cocaine for non-medicinal purposes today.

The cocaine alkaloid was first isolated from the leaves during the 1840s. Such was the interest and so glowing the reports of its effects that coca leaf extracts soon became a common addition to tonic drinks and powders in Europe and America.

The discovery in 1884 of the power of cocaine to deaden nerve endings and its suitability therefore as a local anaesthetic brought about a revolution in surgery. It was found to be the most effective anaesthetic for eye operations, for example, and its use continues for ear, nose and throat surgery today. Awareness of the dangerously addictive properties of cocaine, however, led in time to the synthesis of non-addictive substitutes (such as amethrocaine now used for eye surgery), but without the coca leaf these immensely valuable drugs could not have been produced.

Plants that relax us for the surgeon!

Before certain operations can be performed and to facilitate the surgeon's job, it is important that the muscles to be cut are in a relaxed state. For abdominal surgery and obstetrics, as well as a range of disorders affecting the voluntary muscles, a plant product from the Amazon, made famous as an ingredient of the arrow poison curare, is now widely used as an adjunct to anaesthesia to secure muscle relaxation.

Bolivian woman picking coca leaves

Curare was unknown to the West until the sixteenth century when a group of Spanish conquistadores first recorded its deadly effects. Led by Lieutenant Fransisco Orellana they were, by accident, the first Europeans to travel the entire length of the Amazon river. During their voyage, one of the group was hit in the hand by an arrow and died soon afterwards.

Curare was widely used as an arrow poison at that time by many Amazonian Indian groups and this use continues (though for fewer people) today. Many different sorts of curare are distinguished, each one based on the compounds contained within a different plant or plants. In some cases fruits are used — as with *Ocotea venenosa* used by the Kofán Indians of Colombia and Ecuador, but some of the best publicized ingredients come from species of *Chondrodendron*

Waorani Indians from Ecuador prepare curare for 'poison darts' by scraping the bark of Chondrodendron *or* Strychnos *vines (below) and filtering the pounded bark to extract their potent compounds (left)*

and *Strychnos* vines. The bark of these vines is scraped off in sections and then pounded, and the active ingredients are extracted initially by filtering with cold water. The processes involved are complex and have generally been kept a closely guarded secret by the Indians who use curare. Up to 30 different ingredients can be used for each recipe and recipes vary considerably from group to group. For two centuries the exact contents of curare poison remained a mystery to Western observers and it was not until 1800 that Alexander von Humboldt, who had witnessed the preparation of one kind of curare by Indians of the Orinoco river, gave the first accurate account.

A preliminary understanding of how curare worked on the body arose a little later from the experiments carried out by the eccentric Victorian explorer, Charles Waterton, in 1814. Having already used the black treacly substance in a number of fatal experiments on animals ranging in size from chickens to oxen, he injected a donkey with the drug. Within ten minutes the donkey appeared to be dead. Waterton inserted a pair of bellows into her windpipe through a small hole he had cut for the purpose and started pumping to inflate her lungs. He must have been delighted to see that she soon 'held her head up and looked around' — though he had to keep up his means of artificial respiration for some two hours until the effects of the curare had worn off!

This experiment demonstrated that curare killed by immobilizing the striped (voluntary) muscle tissues so that breathing became impossible. It did not, however, affect the cardiac (involuntary) muscle which meant that as long as breathing was artifically maintained curare could be used as a muscle relaxant. Further experiments carried out on frogs in 1850 by the French physiologist Claude Bernard showed that curare blocked the transmission of nerve impulses to the muscles — taking maximum effect 2 to 5 minutes after intravenous injection. It was not until 1939, however, that the active principle of curare was isolated and not until 1943 that it was introduced successfully into anaesthesiology. Until this time the three main components of anaesthesia, hypnosis, analgesia and muscle relaxation, were all brought about by the use of one anaesthetic agent — usually ether or chloroform. It was only after the patient had entered a deep stage of narcosis that sufficient muscle relaxation was obtained to enable the surgeon to operate easily. The introduction of curare meant that the patient could be kept lightly anaesthetized while obtaining complete muscle relaxation and without the risk of surgical shock.

Today the alkaloid tubocurarine (used intravenously as a clear, colourless solution) is obtained from extracts of the stem of *Chondrodendron tomentosum*, a liana found in the Brazilian and Peruvian rainforest. Its use extends to the treatment of tetanus convulsions, multiple sclerosis, shock therapy and such delicate operations as tonsillectomies and eye surgery.

Plants that hold us together

Once an operation is completed, our wounds could well be helped to heal with more materials from plants. Though the majority of the thousand or so different kinds of fine cord used by surgeons to sew us up are man-made, ranging in strength and texture from stainless steel wire to polyester thread, some stitches are made from linen derived from flax stems, and many are made of silk. Different mulberry species are the most likely food of the silk worms, though several other plants including oak and sal trees (*Shorea* species) and the castor oil plant could also be used depending on the kind of silk-producing worm involved.

Plants that help the heart

A jungle vine from the forests of West Africa, whose extracts have been used like curare, as a local arrow poison, has given us a very valuable medicine for the heart: strophanthin. It acts as a heart stimulant and is particularly useful since it has relatively few gastro-intestinal side-effects.

However, one of the best known and admired of British wild plants, the foxglove, has saved the lives of countless heart attack victims and is helping millions of people who have problems with their hearts to lead normal lives. The foxglove contains digitalis, a mixture of glycosides including digoxin, digitoxin and lanatoside c, which have a powerful effect on the cardio-vascular system. They are used in many cases of cardiac failure since they can strengthen and increase the muscular activity of the heart, stimulating more forceful contractions whilst regulating the heart beat.

FOXGLOVE

Digitalis spp.

Like many valuable medicinal plants the foxglove is highly poisonous. An overdose of digitalis can lead to some very unpleasant symptoms, but in the correct quantities it has enormous value as a cardiotonic, both improving the action of a failing or inefficient heart and reducing a dangerously fast heartbeat.

Although the foxglove has been used for a very long time in folk medicine in Britain and Europe, it was not until the late eighteenth century after the publication of Dr William Withering's scientific investigations into the plant, that its value as a cardiotonic was realized. Withering had spent ten years looking into the properties of the foxglove after finding that an old lady he had been asked to visit had cured herself of dropsy (an unnatural accumulation of fluid in the tissues) by taking a herbal cure containing it. 'This medication' Dr Withering wrote 'was composed of 20 or more different herbs, but it was not very difficult for one conversant in these subjects to perceive that the active herb could be no other than Foxglove'. The improved circulation that had resulted from the woman's ingestion of digitalis had boosted the performance of her kidneys, clearing the accumulated body fluids which are a symptom of the complaint.

Though the common British foxglove *Digitalis purpurea* was the first source of the digitalis glycosides it was later discovered that *D. lanata* the woolly or Austrian foxglove from southeast Europe contained an even greater concentration of these active principles. Today, digitalis is obtained from the dried leaves of both species, but *D. lanata* is the most important source in developed countries. The main producers are the USA, UK, Netherlands, Switzerland, East Germany and the USSR. Some 1500 kg of digoxin and 100 to 200 kg of digitoxin are being used today to treat heart patients each year. Neither of these drugs has yet been synthesized.

Plants fighting cancer

Our continuing struggle to combat the causes and effects of different sorts of cancer has been greatly helped by the strategies adopted by the plant world to help it, too, survive. Nowhere is this more apparent than in the world's rainforests, where plants in competition with each

Korup National Rainforest Park, Cameroon. Rainforests are a vital source of compounds for medicinal and industrial use

PLANT WEAPONS CURING PEOPLE

It is well known that tropical rainforests are the richest and most varied example of life on earth but this variety goes further than the proliferation of physical forms and complex distribution patterns. All plants produce chemicals for a range of different purposes, but the plants of the rainforest are often the richest in defence chemicals, unique biologically active compounds which have evolved to help protect the plants against predators and other damage, but which are not directly essential to growth. Known as secondary chemical compounds or metabolites these substances — most significantly alkaloids and glycosides — have given us the raw materials for some of our most useful and potent drugs.

The bark of trees or vines often contains the highest concentrations of these defence chemicals. A great array of alkaloids and glycosides, as well as fungicidal and bactericidal resins may be present. Two examples relevant to modern medicine are the alkaloids tubocurarine (used as a muscle relaxant in anaesthesia) produced by the bark of *Strychnos* and *Chondrodendron* vines, and quinine, the potent anti-malarial drug present in the root and trunk bark of cinchona trees. The importance of bark as a chemical storehouse is reflected in the pharmacopoeia of one of the

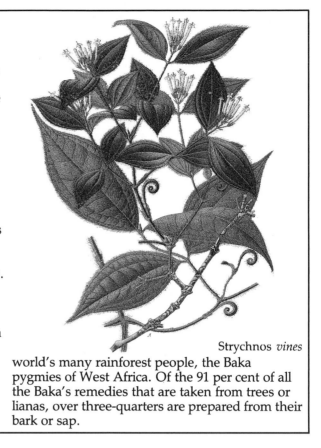

Strychnos *vines*

world's many rainforest people, the Baka pygmies of West Africa. Of the 91 per cent of all the Baka's remedies that are taken from trees or lianas, over three-quarters are prepared from their bark or sap.

other for space, light or nutrients and needing to protect themselves from attack by insects or injury from a host of other hungry forest predators have developed special, often poisonous chemical compounds to help them stay alive.

Those contained in one now very famous forest plant, the rosy periwinkle, have helped greatly in the treatment of certain forms of cancer. *Catharanthus roseus*, also known as the Madagascar periwinkle (after the best publicized of its native habitats), first came to the attention of Western scientists because of the claims made by traditional healers in Madagascar and Jamaica of its potential as a cure for diabetes. Though this diabetic action was not then confirmed, investigations carried out by an American drug company, Eli Lilly, in Indianapolis and by researchers at the University of Western Ontario during the 1950s identified the presence of important alkaloids in the plant. Today over one hundred have been isolated from the roots and leaves.

Two alkaloids in particular, both found in the leaves, are now of major importance as anti-tumour agents. The use of vinblastine (first isolated in 1958) as a substance active against certain cancer cells now means that a person with Hodgkin's disease has an 80 per cent chance of getting better. Other cancers including sarcoma, neuroblastoma and cancers of the breast, lungs and other organs as well as acute and chronic leukaemia are also treated with vinblastine.

Vincristine, which also arrests cell division, works effectively against acute and lymphocytic leukaemia which often attacks children. In 1960, four out of every five children who developed leukaemia died. Now, the use of vincristine has produced a 90 per cent remission rate. Whilst different alkaloids from the rosy periwinkle have a number of other useful effects on the body — they can reduce blood pressure and lower glucose levels in the blood — the use of vincristine and vinblastine has revolutionized the treatment of leukaemia and Hodgkin's disease.

CANCER CURES

Bloodroot (Sanguinaria canadensis)

ROSY PERIWINKLE
Catharanthus roseus

Grown by many people as a small pot-plant with striking pink or white flowers the rosy periwinkle is a member of the poisonous dogbane family and a relative of the hardy, evergreen creeping periwinkles which decorate many British gardens. Present in minute quantities, the rosy periwinkle alkaloids, used as anti-tumour agents, account for just 0.00025 of the dry weight of the leaves. Some 53 tonnes of leaves are required to make 100 grammes of vincristine which is worth around $220 000 per kilogramme.

Many plants have been found to have anti-cancer properties. American mandrake (*Podophyllum peltatum*) otherwise called the mayapple, and the bloodroot (*Sanguinaria canadensis*) were both used by Indian groups in North America to treat tumours and growths of various sorts. A very surprising cancer cure may come, ironically, from the plant that causes some forms of it! The chemical alpha-cembratriene (Alpha CBT) extracted from tobacco leaves by Japanese researchers has been shown to have cancer inhibiting properties when used on mice.

These drugs are currently obtained from plants cultivated in India, Israel and the USA as well as Madagascar. The current annual value of all the alkaloidal materials extracted from the rosy periwinkle has been estimated to be around $100 million.

The rosy periwinkle is not the only plant from tropical forests with a known effect on cancerous cells. At least 2000 other species are believed to have potential as cancer cures, but only one in ten of all tropical forest plants have so far been screened even briefly for their potential. Candidates already of interest include *Tabebuia serratifolia, Jacaranda caucana* and *Croton tiglium* — three tropical trees which each contain unique anti-cancer compounds — as well as species of *Jatropha*.

The continuing isolation of tumour inhibitors from plants, and studies into their structure is yielding a great array of novel drugs, which could be immensely valuable to cancer chemotherapy of the future.

RAINFOREST PHARMACY

The healing potential of the rainforest is enormous, yet Western medicine has only just begun to analyse and evaluate its compounds in any sort of systematic way. For example, only 1 per cent of all the plants estimated to exist in the Amazon have been scrutinized so far. Exciting discoveries are, none the less, occurring all the time. In the Korup National Rainforest Park in Cameroon over 90 chemical compounds have recently been identified with potential for use in both medicine and industry. Thirty-eight of these are new to Western science. But native rainforest people all around the globe have been carefully observing and testing the effects of the plants that surround them for thousands of years. Their own methods of identification and classification are not haphazard and are based on exact empirical knowledge. One group may use over 100 different plants for medicines alone, yet a common misconception still exists that such people need Western medicine to raise them from a state of general malaise and ignorance.

What is true though, is that many health problems have been caused and sometimes exacerbated by the activity and infiltration of non-indigenous people into forest areas. The common cold, influenza, measles and small pox — to which forest dwellers often have no natural resistance and need Western medicines to cure — have killed or weakened hundreds of thousands of them since their introduction from outside.

Other problems such as infestations of parasites have become much worse where formerly nomadic peoples have been obliged to live in groups, whilst the knowledge of effective traditional cures is being lost beneath the pressure to conform. But anti-parasitic compounds — amongst a plethora of other cures — are there in great numbers in the plants. The Amazonian *Ficus anthelmintica* for example, a fig tree which yields a milky latex when cut, was once widely used in the south-west Amazon to combat parasitic worms. In the north-west Amazon Professor R.E. Schultes has recorded over 85 plant species that are used as vermifuges.

In an effort to record and save this information, it has largely been the job of ethnobotanists to study and disseminate the wealth of accumulated local plant lore but a new pride in the validity of these age-old remedies is now emerging in many areas. In the Peruvian Amazon Guillermo Arevalo, a Shaman from the Shipibo ethnic group decided to travel amongst other Indians in the area and train them in some of their lost medical arts. The idea spread and prompted the formation of the AMETRA project (Aplicación de Medicina Tradicional) which is now based in the Department of Madre de Diós. With the full approval of Peru's conventional health authorities, representatives of various Indian communities have been taken on tours of cultivated plots of rainforest plants and their wild counterparts, with instructions as to their identification and use.

Recognition of the knowledge and rights of forest peoples must be considered an essential part of all efforts to conserve and develop what is, after all, their own home and nature's finest pharmacy.

Jacaranda spp., tropical trees which have been found to contain anti-cancer compounds

Rauvolfia *or snake root contains alkaloids which help to reduce high blood pressure*

Plants that lower blood pressure

In Britain alone several million people suffer from high blood pressure or hypertension. In many cases no obvious cause can be found — in some kidney disease, hormonal disorders or a congenital disease is the culprit — but in all cases there is difficulty in pumping blood through the arteries.

One of the most effective modern treatments, helping many millions of people to lead a reasonably normal life comes from a monsoon forest plant, used for at least the last 4000 years by Hindu healers in India to treat snake bites and to cure mental illness. The plant is *Rauvolfia serpentina*, commonly called snake root or rauvolfia.

From more than 24 different alkaloids contained in the plant, and found in the highest concentrations in its roots, the three that have traditionally been most important (rescinnamine, deserpidine and reserpine) are used in medicines to reduce high blood pressure, and as tranquilizing drugs. Although another compound, ajmaline, contained in the roots in fairly large quantities has now become more popular as a hypotensive agent, reserpine in particular has been of enormous importance to medicine.

Since its chemical isolation in 1952 reserpine has been used in small amounts both to lower blood pressure and for the treatment of menstrual and menopausal problems, and in larger doses — still occasionally prescribed today — as one of the most effective plant-based tranquilisers, for the relief of anxiety and to calm the mentally ill. Before its introduction into Western medicine, schizophrenic patients commonly received electric shock therapy or insulin, both of which caused violent reactions.

Despite the long history of its use in Indian medicine it was not until 1954 that the potential of rauvolfia for regulating hypertension was recognized in Britain. By the late 1960s it is estimated that four out of five of all hypotensive drugs prescribed in the USA contained rauvolfia alkaloids.

Reserpine can now be synthesized but the process is a very complex and expensive one and the drug can be extracted much more cheaply from natural sources.

Some 400 to 500 tonnes of rauvolfia roots have been harvested annually in recent years from plants grown in India, Thailand, Bangladesh and Sri Lanka. Unfortunately the great demand for the drug has led to over exploitation of *Rauvolfia serpentina* in the wild and brought it almost to the point of extinction. A sister species, *R. vomitoria*, commonly known as African snake root has now largely replaced it in the world market. This species is much richer in rauvolfia alkaloids than *R. serpentina* and about 800 tonnes of roots are said to be in current use today. These are taken from wild sources on the West African coast, mainly Zaire, as well as Mozambique and Rwanda, chiefly for export to Italy and West Germany. The plant is apparently easy to cultivate and is reported to be used as live fencing, shade for cocoa trees and as a support for the vanilla orchid. Hopefully its fate will not mirror that of *R. serpentina*.

Plants versus mosquitos!

A small, attractive evergreen tree with glossy leaves and fragrant pink or yellow flowers provides the world with one of its most precious medicines — quinine. This alkaloid, stored in the bark of different species of cinchona trees which are native to the humid montane forests of the South American Andes, is the foremost cure for malaria, a disease which affects over 100 million people a year. We now know that the disease is caused by *Plasmodium* parasites transmitted in the saliva of female *Anopheles* mosquitos, but this discovery is relatively recent. The history of quinine's introduction into Western medicine as a cure is a long and complex one.

South American Indians first discovered the effectiveness of cinchona bark for curing fevers and imparted the secret to Jesuit missionaries working in Peru at the beginning of the seventeenth century. In 1633, Father Calancha, a Jesuit priest, wrote of the 'miraculous cures' effected by a powdered bark 'the colour of cinnamon', but it was not until 1639 that the

Cinchona seedlings in a nursery in Ecuador. Quinine is extracted from the bark of mature trees

'Peruvian bark' as it was described, reached Rome and Spain. However, the cure was scorned and rejected by the medical profession who had much to gain financially from treating malaria sufferers inadequately themselves. It was not until a London apothecary, Robert Talbor, successfully cured Charles II of the 'ague' as malaria was then known, at the end of the seventeenth century that Peruvian bark was finally accepted, and the quest began to describe the source of this 'miraculous' new cure.

Joseph de Jussieu, a French botanist who accompanied Charles Marie de La Condamine on the first non-Spanish expedition to enter South America in 1735, was also the first to amass detailed scientific data on cinchona trees, but all his efforts were in vain. In 1761 after nearly thirty years spent searching the Andean mountainsides collecting specimens and writing detailed notes about them and their uses, his entire collection and notes were stolen on the night before he was due to leave the continent for Paris.

Condamine had in the meantime sent a brief description and specimens of cinchona he had seen growing in Southern Ecuador to Linnaeus in 1739, but the live seedlings he intended to send were all washed overboard at the mouth of the Amazon, while being transferred to a larger ship! A further blow to the scientific study of cinchona occurred with the death of José Celestino Mutis in 1808. Physician to the viceroy of what is now Colombia, Mutis spent more than twenty years researching the subject and collecting specimens, but anxious to keep his discoveries to himself, died leaving his papers in a muddle purposely designed to obscure their content.

The first serious study of cinchona species was eventually published in the early part of the nineteenth century by Alexander von Humboldt and the French botanist Aimé Bonpland. After five years of intense study and the collection of a great number of specimens encountered during their exploration of large areas of South America they brought back valuable accurate information on the trees.

In 1820 the quinine alkaloid was isolated from the powdered bark and named by two French doctors (from the Amerindian word for the cinchona tree, quinaquina, which literally means 'bark of barks'). The demand for the drug then rose dramatically. By the mid-nineteenth century reckless exploitation of the wild trees which were cut down so that their root bark could be gathered was threatening the survival of some species.

With malaria rife in their African, Indian and East Indian colonies, the Dutch and British tried to grow their own supplies. The Dutch raised trees in Java and the British, via Kew, in Sri Lanka and India, but the quinine content was extremely low — as little as 2 per cent of the dry bark. Concentration of the alkaloid varies widely between different species and even populations of the same species, and it became urgent to identify and collect seeds from only those trees containing useful quantities of quinine.

An Aymará Indian from Bolivia, Manuel Incra Mamani, came to the rescue, by giving some seeds from a particularly potent stand of trees in Bolivia to the British collector, Charles Ledger who employed him, and Ledger subsequently sent the seeds to Europe. *Cinchona ledgeriana*, as the trees grown from these seeds were named, turned out to contain up to 13 per cent quinine in their bark.

In 1865 Ledger tried to sell his seeds in London. Having already spent considerable sums on collecting seeds which had grown into trees with a very low quinine content (in India), the British Government declined to buy any but the Dutch bought a pound for £6. 7s. From this pound — described by some as the best investment in history — the Dutch successfully grew 12 000 *Cinchona ledgeriana* trees in Java and for almost 100 years controlled nine-tenths of the world's supply of this vital medicine.

The Dutch monopoly was broken by World War II. Though seeds were flown out of Java on the last plane to leave the country before the Japanese arrived in 1942, it was too late to grow the supplies urgently needed for the allied troops stationed and fighting in the tropics. Accordingly a group of botanists were sent by the US Government's Board of Economic

Cinchona spp.

With its bright green shiny leaves, and fragrant flowers hanging in small clusters, Richard Spruce (the Victorian naturalist) decided that cinchona was indeed 'a very handsome tree'.

Native to the montane forests which clothe the eastern slopes of the Andes mountains from Colombia to Bolivia, as well as parts of Northern Paraguay and Brazil, around forty species of cinchona have been described, but four main species are grown commercially today. These are *Cinchona officinalis*, also known as crown or loxa bark; *C. ledgeriana*, grown in Java and India, improved strains of which now yield 16 per cent quinine; *C. succirubra*, and *C. calisaya*, native to Bolivia and southern Peru.

The alkaloids present in the bark of the trees' trunks and roots (which include quinine) are extracted by mixing them in dried, powdered form with solvents such as toluene, amyl alcohol or ether.

Advances in biotechnology have now made it possible to produce quinine from the culture of plant cell tissue. Held in test tubes, the plant cells can be prompted to release quinine into the medium in which they are grown and an absorbent resin added to the culture traps the alkaloid to make extraction simple. Though yields are said to be promising, the process is a very delicate one and costs are high.

ZORN. IC. PL. MED.

Cinchona officinalis. L.

Cinchona officinalis

Cinchona alkaloids

Whilst quinidine, one of the cinchona alkaloids, has become very important in medicine as a cardiac depressant and for regulating abnormalities of the heartbeat, quinine is used (as well as in anti-malarial drugs) in some sunburn preparations and insecticidal treatments. Its best known non-medicinal use, however, is as a bitter in tonic waters and soft drinks. A popular accompaniment to gin, the addition of quinine to tonic water is said to date from the days of the British Raj in India. Advised to take quinine in water for malaria prevention, flavourings were added to enhance the taste and the drink soon became a favourite 'mixer'. Today, whilst 60 per cent of all quinine is used medicinally, the remaining 40 per cent is used by the food and drink industry, the largest end use being as an additive to tonic.

Warfare to Colombia, where they gathered about 12.5 million pounds (5.7 million kg) of dried bark for use throughout the war. Later, new plantations were established in Mexico, Peru and some African countries.

Today most cinchona is grown in Indonesia and Zaire, followed by Tanzania, Kenya, Ruanda, Sri Lanka, Bolivia, Colombia, Costa Rica and India. They produce around 8000 to 10 000 tonnes of bark each year from which some 400 to 500 tonnes of alkaloids, including quinine, are extracted.

In 1944 the first synthetic quinine was made by two American scientists. With the development of antimalarial drugs based on its structure which proved very effective in the control of certain strains of malaria, the demand for natural quinine fell. However, malaria is not a simple disease. Though it was thought to be decreasing, it was discovered in the 1960s that strains of malarial parasite, particularly *Plasmodium falciparum* were becoming resistant to

African guide indicating a medicinal plant (Lavigeria macrocarpa) *used in Cameroon to treat malaria*

these new drugs. There are now areas in the world in which the synthetic drugs (such as Chloroquine, Maloprim and Fansidar) are no longer effective but none of the parasites has yet developed a resistance to natural quinine and other natural cinchona extracts which together have a broad spectrum anti-malarial action.

Quinine is only one of 36 different alkaloids present in cinchona bark. Quinidine, cinchonidine and cinchonine are other very important ones, which have applications in other areas of medicine too.

As malaria continues to evolve, new drugs which incorporate cinchona alkaloids or are based on their structure look set to play a vital part in the race to counteract the deadly work of the mosquito.

Close-up of Lavigeria macrocarpa

Chemicals from plants for AIDS research

> *That many plants not previously considered to be especially useful to man have now been found to contain chemicals which may help us win the war against AIDS should prove to the sceptical that every plant on the planet is worth fighting for.**

At Kings College London in 1981, a member of a research team working on the toxic compounds produced by plants, isolated the chemical castanospermine from the seed of an

*Dr. Linda Fellows, Jodrell Laboratory. R.B.G. Kew, 1988

Australian evergreen tree, the Moreton Bay chestnut (*Castanospermum australe*). The alkaloid turned out to have an unusual structure which had caused it to be 'invisible' in the classic screening procedures used by chemists in the past. Castanospermine is now known to be only one of a series of previously undetected alkaloids which can be considered as 'sugar mimics', since they resemble simple sugars in size and shape.

At the Royal Botanic Gardens, Linda Fellows and her colleagues were studying the distribution of the 'sugar mimic' alkaloids in nature and asking what value they might be to the plants which contain them. They were able to show that many had powerful deterrent or toxic effects on insects and on the digestive systems of insects and mammals. Her group was also collaborating with Dr Stanley Tyms at St Mary's Hospital (London) on the potential anti-viral effects of plant compounds.

When the 'sugar mimics' were tested in 1987 it was found that although they had little effect on many common viruses such as herpes and 'flu, many of them, and castanospermine in particular, had a dramatic effect on the AIDS virus H.I.V. The surface of H.I.V. is covered with proteins to which many sugar chains are attached. These chains play an important role in enabling the virus to attack cells. Castanospermine brings about alterations to these chains and leaves the virus non-infectious.

The 'sugar mimics' have recently been shown to be useful not only in AIDS but also cancer and immunological research, and are providing scientists with a unique 'tool-kit' with which to probe many secrets of biology. The Medical Research Council has awarded a grant to the Royal Botanic Gardens to extend the search for chemicals of this type and many exciting new compounds are already under study.

MORETON BAY CHESTNUT

Castanospermum australe

'That exceedingly ornamental tree of close woods called chestnut at Moreton Bay' as the botanical collector Alan Cunningham described it in 1828, has come to play a part in medicine that could hardly have been dreamt of by the first colonial visitors to tropical Queensland where the tree grows wild.

Though its chestnut-like seeds were eaten by Aborigines after a process of soaking for 8 to 10 days, then drying in the sun and pounding into a coarse flour, in an unprocessed state they are poisonous and yield the valuable if unusual alkaloid castanospermine.

The Moreton Bay chestnut or black bean tree is a native of the rainforests of New South Wales and parts of tropical Queensland. It reaches a height of around forty metres and is an evergreen with glossy dark leaves. The large yellow to orange or red flowers are mostly bird-pollinated. Today the species has become severely depleted through the exploitation of its timber for furniture, inlays and plywood finishes. Its heartwood which is similar to chestnut — beautifully grained, and very strong and durable — was used to make the current Speaker's Chair in London's House of Commons — a gift from the Australian Government to help rectify the bomb damage suffered during World War II.

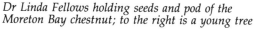

Dr Linda Fellows holding seeds and pod of the Moreton Bay chestnut; to the right is a young tree

Getting around — plants that transport us

Unless you go barefoot, whichever way you travel and whatever your destination, whether the office, a foreign country or the moon, plants will be helping you get there — and back!

When the first space shuttle re-entered the atmosphere after orbiting the earth, cork tiles cut from the bark of the cork oak and used to line the inside of the craft helped insulate its astronauts against the intense heat generated by the descent. Each time Concorde makes a flight its engines make use of oils pressed from coconuts and castor oil beans!

Most of us use plants in less spectacular ways to get us from A to B, but without them, and one in particular, we would find it hard to move at all!

RUBBER FOR ROADS, RAILS AND RUNWAYS

Motor vehicles and aeroplanes

Latex from the rubber tree (*Hevea brasiliensis*) is present in the vast majority of the world's motor vehicle, aeroplane and bicycle tyres, and surrounds the wheels of many of our newest, most efficient trains. This one plant made possible the development of vehicles that could reach their destinations much more quickly, safely and comfortably than before; in fact it revolutionized the potential of every form of transport using wheels.

Though solid rubber tyres for horse-drawn carriages came first, the pneumatic tyre made its successful debut on the wheels of a bicycle. In 1881 John Boyd Dunlop launched his own version of the inflatable tyre (modifying a design that had already been drawn up for the stage coach by the Scotsman Robert Thomson, in 1845) and the painfully bumpy bike ride was no more! With the development of the motor industry at the turn of the century, the tyre had its future assured and became the major product of the rubber manufacturer. This is still the case today, with tyres accounting for 74 per cent of all the uses to which natural rubber is put.

Whilst the first tyres and all rubber goods until World War II were made entirely from a carefully processed base of natural rubber, the Japanese takeover of Malaysian, Indonesian and Thai rubber plantations in 1941 prompted the USA to rapidly develop synthetic substitutes.

The increased demand for car tyres after the War and cheap and plentiful supplies of oil encouraged world-wide expansion of the use of these new kinds of rubbers. Today synthetic polymers such as styrene butadiene (SBR), now the commonest synthetic rubber made from coal and oil by-products, are present in most vehicle tyres in varying proportions, but natural rubber is still a vital component.

Another plant product, rosin — extracted from the trunks of living pine trees, decayed pine stumps and from pulped pine wood — is used in the processing of both natural and synthetic

Rubber from Hevea brasiliensis *trees is used to make tyres for all types of vehicles including Concorde, London buses and trucks!*

polymers to improve tackiness. Alginates from seaweeds, meanwhile, are used to help cream and stabilize the liquid latex.

Different vehicles of course require different kinds and thicknesses of tyre according to the job they must perform. In general, the greater the load carried and the higher the stress, the more natural rubber will be present. This is largely due to two important attributes (which make natural rubber superior to other man-made materials in these respects): its good building tack and strength, and its low heat generation when flexed continually. These properties are especially valuable for large-tyred vehicles carrying heavy loads.

For certain aircraft, including Concorde, where tyres are subjected to tremendous pressures during take-off and landing (and, incidentally, for the space shuttles) natural rubber accounts for 100 per cent of the polymer used. In smaller aircraft the proportion is over 90 per cent of the total polymer. The superior strength of natural rubber, its low heat build-up in service and ability to remain flexible even after long exposure to the sub-zero temperatures encountered in high altitude flying have made it the pre-eminent material for aircraft tyres of all kinds.

Heavy commercial road vehicles such as lorries and buses will also have a very high proportion of natural rubber in the tread of their tyres as well as in the inner layers, where it supplements man-made polymers and steel and textile reinforcements. The proportion of natural rubber overall may be around 75 per cent of the total polymer used. Many off-the-road vehicles too, including tractors, military and logging vehicles, dumper trucks and earth-movers (which use some of the largest tyres in existence) depend on 100 per cent natural rubber. Since the problem of heat generation is not so severe with cars (with their smaller wheels and lighter loads), the external tread of their tyres is usually made almost entirely of synthetic rubber, such as styrene butadiene, which wears well and gives good grip on roads.

Tyres are complex composite structures, made up of a number of different components which include — besides the tread — the carcass, sidewalls and inner liner.

The most common type of tyre for passenger cars today is the radial, in which the tread is braced by steel cords or layers of synthetic fabric. Tyres of this kind, especially all-weather radials, may contain as much as 50 per cent natural rubber, chiefly in the inner casing components and sidewalls. Natural rubber is vital to their manufacture as its properties of strength and 'tack' (adhesion) enable the shape of the tyre to be maintained while it is being built up. Cross ply tyres, which have generally been replaced world-wide by radials, used considerably less natural rubber in their construction.

RUBBER — FROM TREE TO TYRE

The source of the natural rubber indispensable to tyre manufacturers is the rubber tree *Hevea brasiliensis*. Though native to Brazil, most of the world's rubber supply now comes from trees grown in commercial plantations and smallholdings in Malaysia, Indonesia and Thailand.

When the bark is cut, often in an oblique groove round part of the trunk, an interconnecting system of tubular vessels just beneath its surface exudes rubber latex, a thick milky liquid, that is in fact a suspension of rubber particles in a watery fluid. Tapping is a highly skilled operation since the best latex-yielding vessels, which occur in the tree's inner bark are nearest the cambium growth layer. If damaged, regeneration of the bark is hindered and the tree's life shortened.

A remarkable feature of the rubber tree, however, is its ability to renew within a day the supply of latex in the area of bark just tapped, after the initial flow (which lasts for 3 to 4 hours) has stopped. By cutting a thin sliver of bark from the lower edge of the groove the process can be repeated many times during the 20 to 25 years that the tree remains useful.

A more modern method is to puncture several small holes in the bark in place of a continuous cut. This reduces the area of bark that needs to be used but the incisions must be treated chemically to stimulate a sufficient flow of latex.

The latex is caught in small cups attached to the trunk by rubber tappers. When the cups are full they are emptied into buckets and the latex is strained to remove any impurities. After being transferred to large tanks, acid is added to coagulate the latex, clumping the rubber particles together. As it dries, the mixture is divided into thick sheets by slotting metal plates into the tank.

After any water present has been pressed out, these sections are dried and sold as either 'ribbed smoked' or 'air-dried' sheets. Increasingly the coagulum is crumbled instead by adding a small amount of castor oil and passing it through a series of rollers. Crumb rubber is thoroughly washed and dried in hot air ovens before being pressed into standard bales and packed for export.

Tyre-making is a complex process since both radial and cross ply types comprise many layers that are built up step by step. Production begins by masticating the raw rubber: working it to soften it and to prepare it for mixing with other chemicals. Early on in the production of rubber goods at the beginning of the nineteenth century it had been found that raw rubber was an awkward material to process and shape *(continued)*

Rubber latex being tapped

Hevea brasiliensis

Acid coagulates the rubber latex into clumps (above) then the mixture is divided into thick sheets by slotting metal plates into the tank. Any water present is then pressed out (below)

satisfactorily, and that a means of softening it was needed. Though mastication (invented by Thomas Hancock in 1820) solved this problem, and enabled rubber to be used to make such innovative products as the waterproof raincoat, it had the inconvenient attribute of making the rubber mixture stiff when cold and soft and sticky when hot!

By adding sulphur to the rubber and heating the two together, the chemical linkages between the rubber molecules were stabilized and the problem was solved. A compound was thus made that was elastic and relatively temperature insensitive. Invented by Charles Goodyear in 1839, the process, known as vulcanization (after Vulcan, the Roman god of fire) is essential for the production of all modern rubber products, including tyres.

For tyre manufacture, other chemicals are added to the rubber base before vulcanization, both to assist in the process and to protect the rubber from the effects of heat, light and ozone in the air. Special fillers are also mixed in to give strength and stiffness. The commonest, carbon black, turns the pale or dark brown rubber to the familiar black of tyres.

The various components of the tyre, its tread, sidewalls, carcass and other smaller items are shaped separately from the rubber and chemical mix, or from synthetic rubber, and assembled in a complex building operation. Vulcanization is then carried out by heating the components in special moulds which also develop individual features such as the tread pattern.

THE PAST PRICE OF RUBBER

Like many other plant products that we take for granted today, the potential of rubber was first discovered by Amerindian peoples of South and Central America.

A famous game once played by Mayan Indians revolved around the hitting of a heavy rubber ball — using only the head, legs or shoulders — through a stone ring fixed to the side walls of a special court. Columbus found smaller rubber balls being used as toys by children in Haiti and on his return to Spain presented several to Queen Isabella. A little later, in the early sixteenth century, Cortés' men in Mexico used rubber latex to waterproof their clothing, having observed the use made of it by local people. The Omagua Indians, meanwhile, who populated the banks of the Central Amazon in large numbers until the arrival of Europeans had long made pouches, bags, footwear and even syringes from natural rubber.

Though the substance had intrigued those Europeans who encountered it, the scientific world was not really interested until Charles Marie de la Condamine returned with samples in 1744. Commercial development did not begin until 1823, when Macintosh took out his patent for various kinds of rubber-proofed fabrics. The most significant impetus however, was Dunlop's patent for 'a hollow tyre or tube made of India-rubber and cloth, or other suitable materials', which precipitated what had been a moderate but profitable industry into a wild commodity boom.

From about 1880 to 1911, with the spiralling demand for rubber for the motor tyre industry, a fever unparalleled in its barbarity and greed gripped certain individuals who set themselves up to 'manage' for their companies the tapping of rubber latex from Amazonian (*Hevea brasiliensis*) trees. Using Barbadian slave labour to oversee their workforce, thousands of Amerindians were forced, using horrific cruelty to gather the precious latex. This cruelty resulted in the deaths of thousands of people: some 40 000 Indians from the Putumayo region alone were killed in six years.

By 1914 the great fortunes that had been made at the expense of the native people of the Amazon had collapsed. The Brazilian seeds taken to South East Asia in the late 1870s had prospered and their yield of latex had exceeded that from wild South American trees for the first time. The Amazon rubber boom was mercifully at an end.

THE SEEDS OF WORLD RUBBER

Eager to have their own rubber plantations in South East Asia and to break the Brazilian monopoly on this lucrative raw material, the British tried for some time to grow *Hevea brasiliensis* seeds and plants in England and abroad. (continued)

Hevea brasiliensis *seedling*　　*India rubber tree* (Ficus elastica)　　*Ceará rubber tree* (Manihot glaziovii)

All these attempts met with failure until a batch of 70 000 seeds, carefully packed in banana skins, were shipped from Brazil by Henry Wickham, arriving at the Royal Botanic Gardens, Kew in 1876. Only 2397 of these seeds germinated successfully (in Kew's orchid greenhouses) and most of these were sent to Sri Lanka for planting. It was in Malaysia, however, that the Kew seedlings were to assume greater economic importance, following the development of rubber tapping techniques in Singapore. From just 22 seedlings grown here are descended the 2 million hectares of rubber trees growing in Malaysia today. Malaysia is now the world's biggest supplier, producing over 1.5 million tonnes of rubber each year, whilst Indonesia produces 1.2 million tonnes and Thailand almost 1 million. Total world production is in the region of 5 million tonnes.

Careful selection and breeding techniques as well as the introduction of chemicals which stimulate the flow of latex from the trees have greatly increased output over the years, although genetic material from wild trees native to the forests of the Amazon remains particularly important. In recent years *Hevea brasiliensis* seedlings from the Brazilian states of Acre, Rondonia and Mato Grosso have been highly rated in terms of their potential latex yields.

RUBBER FROM OTHER PLANTS

Hevea brasiliensis is not the only plant that yields a useful rubber latex. Many other species of plant and even some fungi have been found to contain it, and until the end of the nineteenth century several were in use. They include the Ceará rubber tree (*Manihot glaziovii*), from Brazil, several species of *Castilla* tree, especially *Castilla elastica* which caused a short-lived rubber boom in Panama and Mexico, and the India rubber tree (*Ficus elastica*), now commonly grown in Europe as a houseplant. In West Africa, *Landolphia* vines, and *Funtumia elastica* were of particular importance and reports of their potential were part of the impetus behind the Belgian take-over of the Congo in 1883. Even a dandelion (*Taraxacum bicorne*) supplied rubber for a while, providing the Russians with latex during World War II.

Though *Hevea brasiliensis* is the only source of commercial rubber used today, guayule (*Parthenium argentatum*) a shrub native to northern Mexico, Texas and southern California, may have great potential for the future. In 1910 it supplied 10 per cent of the world's natural rubber and when supplies of Malaysian and South East Asian rubber were cut during World War II, large investments were ploughed into guayule cultivation in the USA. Though these efforts were abandoned when the War ended the plant has once more come under close investigation in North America. One of its major advantages is that it can be grown in very arid regions, and as the demand for natural rubber from conventional rubber trees is predicted to outstrip supply, guayule could become a valuable alternative. The US Airforce are already very interested in the prospect of guayule rubber making up at least a proportion of the rubber in their aircrafts' tyres.

This monorail runs on rubber tyres

Tyres are not the only parts of motor vehicles that depend on this versatile, long-lasting material. Its flexibility and water-tight properties have made it ideal in the past for engine radiator hoses and the weather seals around doors and windows. Natural rubber plays an important part in interior suspension too. All engines are mounted on rubber compounds to cut down vibration and noise and many small components such as bushes and gaskets hidden away under the bonnet or in the chassis of the car, as well as mats for our feet, use natural rubber.

And there is rubber under the road as well. The bearings of many road and rail bridges are steel-laminated blocks of natural rubber, which allow the structures to expand and contract as the temperature goes up and down.

High speed trains on rubber!

Astronauts, road users and plane passengers are not the only travellers to benefit from the versatile rubber plant. The tyres of the trains which speed along the Paris, Mexico and Montreal Metros, as well as several transit systems in Japan, are also made in part from natural rubber. The percentage is quite high, about the same as that for commercial lorries, accompanying synthetic polymers and steel and textile reinforcements.

Although the trains are not the fastest in the world, their rubber tyres enable them to accelerate and decelerate quickly and to tackle gradients as steep as 1 in 7, reduce noise levels considerably, and lessen vibrations, making them amongst the quietest and most comfortable to travel in. Sapporo, the largest city on the northern Japanese island of Hokkaido, and seven other major cities in Japan now have subway trains that run on tyres. Several of the country's commuter trains and monorails like those of Tokyo and Kyushu that run above the ground are similarly equipped and many new tyre-based transit systems are expected to be in operation by 1995.

The French however, who now have four metro systems that run on rubber (Paris, Lyons, Marseilles and Lille), pioneered this quieter, more comfortable form of rail travel. After Andre Michelin had spent a far from restful night on an express train thundering its way to the Côte d'Azur in 1929, he asked his brother Edouard to try and find a way of improving this most noisy means of travel. Edouard's answer to the problem was the pneumatic tyre! By 1931 the first Micheline, a commercial railcar prototype had been unveiled. The new vehicle, powered by a petrol engine, looked very much like a lorry but had tyres specially adapted to run along thin rails. As designs progressed, however, vehicles came to look much more like modern trains.

Running on twenty-four pneumatic tyres — and with rubber used to mount the engine and improve the suspension system — the rail cars were a great success. They were able to travel fast and quietly to please the passengers and they also cut down railway maintenance. In 1948 Michelin introduced a complete rubber-tyred train designed to be pulled by a steam locomotive and three of these plied the route between Paris and Strasbourg. But the discovery that quietness in train travel could be achieved more easily by welding rails together into longer lengths and by using rubber pads, led to the discontinuation of the designs.

However, rubber tyres could also be used in other ways. In 1951 the Paris Metro became the first exponent of a brand new design by Michelin which enabled ordinary tyres to run on a simple concrete track, with steel wheels and rails to provide guidance and ensure safety in the event of tyre failure. The great success of the Paris Metro led directly to the building of the Montreal and Mexico systems which now carry millions of passengers each year.

Though London's underground does not run on rubber tyres, natural latex still plays an important part in the railway's functioning. Whilst rubber blocks help support the track, rubber springs cushion the bodywork of the carriages. Escalator handrails are also made in part from natural rubber.

Escalator handrails on the London Underground are made in part from natural rubber

London Underground trains have rubber springs cushioning the bodywork and flooring made of rock maple (Acer saccharum)

TREES FOR TRANSPORT

Wood has been indispensable to the development of rail, road, air and sea travel all around the world. Though its use is now reduced considerably in many forms of transport, it is still vital for a number of construction purposes.

Rail travel

To date, the flooring of all London Underground's trains has been made of rock maple (*Acer saccharum*) imported from Canada. Though the floors of new rolling stock are expected to consist of man-made materials, all of the carriages in current use rely on maple wood and many of them display it in their distinctive grooved flooring. Beneath the train, shoe beams and boards supporting the collector shoes (which pick up the 600 volt DC current) are laminated with one of several tropical hardwoods.

ROCK MAPLE

Acer saccharum

Known in Canada as the sugar maple (the source of maple syrup), in the USA as white maple and classified as 'hard maple' in the timber trade, this tree is a native of north temperate regions and produces pale, normally straight-grained timber which sometimes carries a distinctive 'bird's eye' figure. The wood is slightly heavier than beech and of high density. It is very resistant to abrasion and the surface wears smoothly without disintegrating. For these reasons, maple timber has been much used to make the flooring of industrial buildings, dance halls, squash courts, bowling alleys and gymnasiums, as well as underground trains. Rock maple was also used by London Underground until recently for escalator treads.

The Paris Metro actually runs on a track made of the West African rainforest timber ekki. This outstanding wood, shipped mainly from Cameroon, comes from one of the largest African trees, *Lophira alata*, which can grow to 60 metres in height. Its clear straight bole makes it suitable for cutting into very long timber lengths. Dark red-brown or purple-brown in appearance, with fine whitish flecks, ekki is 50 per cent heavier than oak and extremely strong. Its exceptional density (similar to greenheart) means that it cannot be nailed or fastened unless the wood has first been drilled, but this makes it extremely durable, and resistant to fungi and wood-boring insects.

Ekki has also been used in Europe alongside several other tropical hardwoods such as afzelia (*Afzelia bipindensis; A. pachyloba*) from Africa to build the underframes and bottom cladding of railway vehicles. Other species used for railway sleepers (although not in Britain) include kempas (*Koompassia malaccensis*) from South East Asia and mora (*Mora excelsa*) from South America. Though many have now been replaced with concrete, oak was in the past the principal timber for sleepers used by British Rail and a number are still in use today.

Inside the passenger carriages on British railways, decorative veneers of teak, sapele, walnut, cherry, box and ash were once very common, but have now mostly been replaced by man-made composite materials.

Wood on the road

Road vehicles still rely on wood to varying degrees. Though the frames and bodywork of modern cars are now generally made of steel, aluminium or fibre glass, the British Morris Minor and Morris Mini Travellers and many of the older European and American 'station wagon' vehicles have wooden frames. They are joined in some developing countries by a large number of trucks and buses whose bodywork develops a certain charm with age.

Though the last new Morris Traveller was built in 1971, around 35 000 are still on the road. Along with almost any other renovation work required, these cars can still have their wooden frames and interior fittings repaired at special centres. The wood used for the Traveller's rear frame and roof rails is ash. It is one of Britain's toughest native hardwoods and is well suited to frame-making, since it bends easily when steamed and absorbs shock very well. The creamy golden colour of the timber is certainly very attractive with a surface grain that resembles the contour markings on a map.

The Morris Traveller has an ash frame

The frame of the 'Wanderer', an early caravan, was made mainly from hardwoods

Caravan frames were once made extensively from tropical hardwoods, such as Keruing (*Dipterocarpus* spp.) and Mersawa (*Anisoptera* spp.) from South East Asia. Today most caravans are made externally of aluminium, with plywood panelling on the inside to give shape to the roof and form the floor. The plies are generally cut from softwoods, with small pieces of solid timber used to frame windows and doors.

The floors of all London buses are currently made of strong 'marine' plywood — overlain with man-made coverings — since it has been found that this material is both simpler and cheaper to install than aluminium or similar alternatives. The main timber used for facing the plywood is birch grown in Finland though softwoods grown elsewhere in Scandinavia and Northern Europe are likely to be used to make the other plies. London buses also have seat bases and backing boards made from plywood panels.

Other wood-based boards are very important for the construction of car interiors. Perforated hardboard for instance, lines the roofs, boots and door panels of many modern makes of car. The trees involved are likely to include mixed conifers such as pine and spruce species and various hardwoods, especially eucalyptus if the boards originate in Australia or Brazil.

Perhaps the most familiar item to be made of wood in many cars is the dashboard or fascia, and in certain more luxurious models, the doorcappings (or waistrails), picnic tables (fitted to the rear of the front passenger seats) and centre consoles.

In most cases thin veneers of woods selected for the beauty of their grain and markings overlie a framework made of plywood or solid timber, carefully cut and assembled by hand. Though other woods are also used — elm for example in some of the more expensive cars — the pre-eminent material for decorative veneers is walnut. A particular kind of walnut wood is especially favoured. An irritation or injury to the tree causes it to form what is known as a burr on its branches or on the side of the trunk, sometimes extending beneath the ground. When cut or 'peeled' carefully by machine, a beautiful streaked and mottled pattern is revealed, varying with every burr — a figuring that has been highly prized for centuries by cabinet makers and all those working with decorative woods.

Rolls-Royce and Bentley cars make particularly important use of burr walnut today. Each year 6500 square metres of this fine wood is fitted to their interiors adding significantly to the style and elegance of each car. A typical fascia will comprise eight separate pieces of veneer, but these are aligned and matched so carefully that they give the appearance of one solid piece

*English walnut (*Juglans
regia*)*
*Black walnut (*Juglans nigra*)*

WALNUT

Juglans spp.

The English or common walnut (*Juglans regia*) is one of fifteen walnut species native to various parts of Asia and North and South America. Despite its name, the natural range of English walnut extends from the Himalayas through Iran, the Lebanon and Asia Minor into Greece, and the tree was not introduced to Britain until the middle of the fifteenth century. Heavy cutting since that time has greatly reduced the availability of the wood however and commercial supplies have come mainly in recent years from France, Turkey, Italy and Yugloslavia. The tree reaches a height of between 24 and 30 metres with a diameter of 1 to 3 metres. Its edible nuts, each enclosed in a wrinkled, light brown skin, form inside plum-sized green fruits and are as sought after as the tree's timber.

Though the sapwood is a pale straw colour and clearly defined from the heartwood which is generally greyish-brown, walnut timber varies considerably in appearance according to local conditions. Infiltrations of colouring matter — usually an attractive smokey-brown or reddish-brown — produce darker streaky patterns in the heartwood, and this decorative effect is often accentuated by the naturally wavy grain. Whilst Italian walnut wood tends to have darker more elaborate markings than that from France and Turkey, the markings on these woods are paler than those of English walnut. The best quality walnut was said at one time to come from Ancona in Italy, and this name is still used today to designate any well-figured, dark, streaky walnut.

Also highly sought after for its timber is the black walnut (*Juglans nigra*) native to North America, and to a lesser extent the butter-nut (*J. cinerea*) and the Japanese walnut (*J. ailantifolia*).

Whilst common and black walnut timbers are often associated with furniture of the Queen Anne period, they continue to be used for decorative veneers and panelling and are also used in solid form to make the butts and stocks of guns and rifles. Burr walnut is especially sought after for veneers. Cut from irregular outgrowths which form on the trunks of many trees, the wood has a beautifully mottled pattern and is currently in great demand for the fascia panels and other interior fittings of luxury cars.

Various unrelated timbers are sometimes described as walnut, including satin walnut from the USA, African walnut, East Indian walnut or kokko from the Andaman Islands and Queensland walnut from Australia. However, only *Juglans* species are true walnuts.

The interiors of Rolls-Royce cars are embellished with walnut veneer

of wood. Each section of veneer is in fact split in half and opened out to give a mirror image of the burr patterning. Veneers cut from the same log will be used for sections of the waistrails, centre consoles and picnic tables to harmonize the car interior as a whole.

Different walnut species are used selectively for different effects. The principal one in Rolls-Royce models is usually black walnut (*Juglans nigra*) grown in California which is used as a burr veneer for the upper surfaces of tables and centre consoles, and in solid, straightgrained form for the tops of the waistrails which are often stained a darker colour. Straightgrained European walnut (*Juglans regia*) is used for facias and waistrails in some Bentley models. Other woods are also in evidence. The lighter 'Australian striped walnut', not in fact a walnut species but *Endiandra palmerstonii*, is used for decorative cross-banding in Rolls-Royce models and bird's eye maple (*Acer* spp.) is also sometimes chosen for the various surfaces. All of these acquire their impressive glass-like sheen and finish after careful coating with a primer and several layers of polyester resin lacquer before sanding down and polishing.

Other woods are involved again in the structures underlying the veneers. For the sides of the waistrails as well as glove box lid and matching end panels, solid West African mahogany is typically used. The fascia is made of birch-faced plywood, overlaid with a panel of the West African hardwood makoré (*Tieghemella heckelii*) on the back and straight grain walnut on the front to prevent it twisting. For its great strength and considerable resistance to warping and twisting, birch-faced plywood is used elsewhere in the cars under much of the veneer. As a structural timber, long valued for its strength and ability to be curved without splitting, ash is used in the hood structure of the Rolls-Royce Corniche and Bentley Continental.

All the wood used for these sumptuous cars is carefully checked for quality and moisture content before assembly. The sheets of walnut veneer must be soaked in water, then flattened in a press before air-drying for four days, to reach a moisture level of 10 per cent or less. Every detail is inspected. Even minor flaws in the natural figuring are corrected with a paintbrush and matching pigment!

Sweet chestnut (Castanea sativa) provides tannin, used to strengthen leather upholstery

Trees for leather seats.

Leather car upholstery also owes part of its distinctive quality to plants. Though, like most soft leathers used today for clothing or shoe uppers, the cattle hides made into seats are tanned with chromium to give them suppleness, they are also immersed in plant tannin solutions to give them strength and to stop the leather from stretching with constant use. The most likely sources of plant tannins used today in Europe, North America and Japan are black wattle trees (*Acacia mollissima*) grown chiefly in Southern Africa, quebracho (*Schinopsis* spp.) from Paraguay and Argentina and sweet chestnut trees (*Castanea sativa*) from Italy and France.

Polish from plants

To leave the car in sparkling condition plants can help polish up the paintwork. Carnauba wax which forms on the young leaflets of the Brazilian wax palm (*Copernica prunifera*) is a component of many car polishes today. Only small amounts are needed, as the wax is very hard, but it is still one of the best materials for the job, supplementing the larger quantities of oil-based paraffin wax and small amounts of beeswax which with various solvents make up the polish base. Sodium alginate from seaweeds is also sometimes present in the polish, since it has valuable suspending properties and helps bind all the other ingredients together.

Leonardo da Vinci's Ornithopter

The Wright Flyer

TREES FOR AIR TRAVEL

As a light but strong material, easily worked and in plentiful supply, wood was the natural choice for the builders of the world's early aircraft, and is still used today for certain kinds of light aeroplane and glider. Leonardo da Vinci's man-powered Ornithopters, for which several designs were sketched in the late fifteenth century, were to have flown by means of bird-like wings (ingeniously controlled by human hands and feet) made of a wooden frame (probably beech) covered with a lightweight fabric. Unfortunately, Leonardo, who spent many years perfecting his delightful 'flapping machines', misunderstood, like the few who had preceeded him and most of those to follow for the next 300 years, the basic principles of flying.

By the early nineteenth century, however, as aerodynamic principles were beginning to be understood, experiments with kites and gliders made of lightweight wood and fabric brought the reality of aviation within reach. In 1853 the first successful full-scale glider, designed by Sir George Cayley and carrying his coachman as a passenger, made its historic maiden flight.

Complaining that he had been 'hired to drive and not to fly', the coachman who stepped out afterwards promptly gave in his notice, unaware or unconcerned that he had just made history! Further experimentation by many dedicated individuals and the perseverance of the Wright Brothers, at last brought the first powered take-off, by the Wright Flyer, in 1903. The materials used to make the framework and vertical and transverse struts as well as the wing spars of many of these early aircraft — chiefly bi-planes up to 1935 — were ash or spruce, or sometimes bamboo lengths. The whole structure was assembled by gluing, pinning or bolting the component parts together, and was braced with numerous tension wires. The wings were covered with strong fabric, such as varnished silk, unbleached muslin or rubberized, water-proof linen and had ribs of plywood, generally birch faced, a very versatile material that was also used to panel the fuselage where necessary. Steel tubing and small amounts of aluminium were also used to help strengthen the aircraft. For lightness seats were made of wickerwork (woven willow).

In 1912–13 the French-built Deperdussin racers introduced the monocoque design for aeroplanes. For the first time the stresses were carried by a single outer shell obviating the need for any internal skeleton. Their streamlined body was made from three layers of tulip wood veneer (*Dalbergia cearensis*) glued together over a strong hickory frame (*Carya* spp.). Though these particular materials proved too expensive to be commonly used, the great strength and flexibility of other woods and wood veneers, which could be curved when heated, was essential for the early monocoque designs.

From the 1920s onwards, military aircraft in particular began to make much more use of metal for their construction, using an all-metal 'stressed skin' in place of wood. But World War II brought a dramatic revival of this now somewhat maligned material mainly because of the need to conserve scarce and valuable aluminium resources. The famous British combat aircraft, the Mosquito — made almost entirely of wood — turned out to be one of the most successful aircraft of all time. A sophisticated mix of hard and softwoods, the Mosquito's stream-lined shell was made of balsa wood (*Ochroma lagopus*) sandwiched between two birch veneers. This balsa 'sandwich' was extremely stable and needed no additional stiffening. The torsional strength of the plane's rear section was cleverly achieved by laying the plies diagonally so that their grains crossed at right angles round the frame. The wings also had

The largest flying boat ever built, the 'Spruce Goose', is currently on display at Long Beach, California. Designed during World War II for the Navy, the War was over before the boat had been completed. Millionaire Howard Hughes stepped in to rescue it, at a cost of $22 million. It made only one short flight in 1947 before being put in store.

As its name suggests, the timbers used to make the flying boat were largely spruce (*Picea* spp.).

The Spruce Goose

One of the de Havilland Mosquitos

This is one of the 7781 de Havilland Mosquitos built for combat during World War II. Made entirely of wood, its frame comprises two load-bearing skins of birch, bonded to a low density balsa core. Spruce wood has been used for the wing members.

inner and outer skins of plywood, fixed by strong formaldehyde cement to stringers made of spruce. Even though the heat and damp of tropical environments were to cause problems with the wood adhesives, two important advantages of the Mosquito's construction were that it was much less easy to detect by radar, and if damage was sustained, its panelled sections made of wood were far easier to replace than metal.

Today some light aircraft and gliders are still made of wood — especially those built by individual enthusiasts. Despite the competition from steel or aluminium-framed planes, many of which have a bodywork of composite materials such as glass reinforced plastic or moulded carbon fibre, two kinds of wood construction are in use: the box frame, often made of sitka spruce (*Picea sitchensis*) covered with sheets of plywood (plies) or strong fabric; and the semi-monocoque, built from plywood moulded into shape, and with internal structures made from other timbers. Birch and mahogany are currently the chief woods used for facing the plies.

Mahogany also played an important part in early aviation. Propellers were often made entirely from this wood, since for its weight its strength properties are extremely good. Central American mahogany was chiefly used, cut into a number of laminates which were glued together and shaped by hand to form the blades. After 1935 'densified' wood was generally adopted, however, using other timbers. This was made by compressing multiple laminates together to double their density. Today some small aircraft still use wooden propellers, chiefly mahogany and other tropical hardwoods.

PLANTS ON WATER

Totora reed boats, made on the northern coast of Peru

Totora reeds

About 500 km from Lima at Huanchaco on the northern coast of Peru, local fishermen still ride out to sea on 'caballitos de totora', their 'little horses' made from reeds. Identical craft, distinguished by their graceful curving prows which dip and bob over the rolling Pacific breakers, were being used by the Moche people in pre-Inca times. Later, in the sixteenth century Spanish chroniclers (who first described the boats as 'little horses'), marvelled at the fishermen's prowess: 'everyone set on horse-back, cutting the waves of the sea, which in their place of fishing are great and furious'.

TREE TRUNKS FOR BOATS

It is easy to imagine that the first successful means of water transport was developed simply by observing the ability of trees to float and their imperviousness to water.

Today the trunks of many large forest trees which have been skilfully hollowed out and further shaped and hardened with the help of fire and steel instruments serve thousands of communities of sea, lake or river-dwelling peoples.

In West Africa for example, the trunk of the large forest tree iroko (*Chlorophora excelsa*) is shaped and hollowed out by repeated charring then adzing to make a sturdy dug-out canoe. In Polynesia, breadfruit trees (*Artocarpus* spp.) provide good boat-building timber that is easily worked. Palm species and mango trees are also useful for making hulls, whilst coconut fibres are sometimes utilized for lashing outrigger poles (often sections of bamboo) into place.

A South American hardwood, distinguished by its extraordinary softness and lightness — balsa (*Ochroma lagopus*) — has given us a timber much used in the past for raft-making by South and Central American Indians and still in use today in certain areas by fishermen. It was a sturdy ocean-going balsa raft fitted with cotton sails and carrying a precious cargo which included gold and silver jewellery that gave gold-hungry Spaniards sailing off the coast of Ecuador their first evidence of the advanced civilization and fabulous wealth that lay before them in Peru.

Thor Heyerdahl's historic 5000 mile (8000 km) voyage from Peru to the Tuamato Islands near Tahiti in 1947 was also made on a balsa raft, the 'Kon Tiki'.

Nowadays balsa wood has found a different but none the less important use in modern shipping. As the lightest of all commercial timbers — noted for its very low thermal conductivity and high sound absorption — it is used in large quantities as an insulation material in ships carrying liquid gas and other goods that need similar cold storage. Balsa's extreme buoyancy has also made it suitable for the manufacture of life-saving equipment such as floats and buoys.

BAMBOO FOR BOATS

Bamboo has been of very great importance to the boat-builders of China. The sectioned culms of the bamboo plant (formed by nodes which divide the stems into separate compartments) are said to have given the Chinese the idea of making water-tight bulkheads, and these were indeed fitted to their junks 2000 years before they appeared in the West. Today the huge variety of junks, sampans and rafts that navigate the waterways of China carrying every sort of cargo are made from a

Making a dugout canoe in Belize

range of different timbers, but bamboo features prominently in many. Besides being woven into screens, roofing mats and deckhousing, split lengths are peeled and their fibres made into ropes and cables of great strength, resistant to rot and stretching and very light to handle. Bamboo is also used to make rods and laths that help hold sails to the masts of certain boats.

In Formosa, the stems of the giant bamboos *Dendrocalamus giganteus* and *D. strictus* are fashioned into light sea-going trapezoidal rafts, which curve up sharply at the prow and less so at the stern.

Dugout canoes in the Peruvian Amazon

Sampans in Singapore

To make the boats, totora reeds (*Schoenoplectus riparius*) grown in the vicinity are cut, dried and lashed together in two thick bundles, about 3.5 metres long. These are then tied side on to form the raft-like, buoyant base, with the rear end blunt and the front end curving to a sharply tapered point. The fishermen kneel or sit astride the caballitos, and use a paddle to help manoeuvre them out to sea.

Some 1200 km to the south, a different species of totora reeds (*Schoenoplectus tatora*) have been used by the Uros and Aymará Indians of Lake Titicaca for centuries to make elegant boats, for travelling, fishing, hunting ducks and carrying supplies. The reeds, which grow at the edges of the lake and are also cut by the Uros to form the 'floating' islands on which they live, are stacked up to dry and then lashed together with ichu (*Stipa ichu*) a tough bushy grass that characterizes the altiplano (the surrounding treeless, windswept plain). Four long bundles which taper at both ends are constructed and tied firmly together to make the distinctive shape of the boats. Sails too, are sometimes made from totora reeds which have been split and sewn together lengthwise, but the Indians generally propel and steer their boats with the help of long poles. As the reeds begin to rot and disintegrate after about 6 months in the water, new ones are cut to replace them.

Reeds and reedlike plants have been used by various peoples throughout history, especially where trees are scarce, to make rafts and simple vessels. In Ethiopia today, the people of Lake Tana cut papyrus plants (*Cyperus papyrus*) and tie them into bundles to make boats. But much more durable raw materials for water transport have come almost universally from trees. From the bark canoes of North American Indians to the sturdy hulls and framework of medieval sailing ships; from Phoenician merchantmen and Roman quinqueremes to Britain's 'clippers' and from balsa rafts to modern lifeboats, trees have provided most of the essential raw materials for sea and river-going craft.

Naturally, a great variety of different timbers — and one exceptional grass, bamboo — have been used around the world by different peoples for boat construction depending on the type of craft and its function. In Northern Europe, from at least the time of the Vikings until the nineteenth century, the great merchant ships and naval vessels depended largely on oak for their strength and durability. For many of the larger ships, whose often extraordinary voyages made possible the expansion of western civilization — carrying cargoes of slaves, and passengers to and from the distant lands they found — huge single timbers, such as those which formed the stern posts, were much in demand to take the ships' tremendous loads and stresses, though small, overlapped and interlocking pieces were often substituted. Curved members such as ribs were similarly often formed from individual sections fitted together, though single pieces taken from the naturally occurring curves in certain trees were preferred. Such timbers could only be cut from enormous ancient oaks, and as the size of fleets increased, the difficulty in finding suitable trees became more difficult.

The building of a Tudor warship like the *Mary Rose* — which sank in 1545 — was a major undertaking both in terms of man-power and materials used, and required an estimated 14.5 hectares of trees in southern England. For an eighteenth century gunship like the *Victory*, it is thought that about 32 hectares of oak forest would have been cut. Frames, deck beams, flooring, side bracing and supports as well as out-board planking were all frequently made of this resilient timber.

Though the British Navy had a guaranteed supply from the Royal forests — chiefly the New Forest in Hampshire and the Forest of Dean — most dockyards still had to obtain much of their timber from contractors who monopolized the supply. The whole system became corrupt and constant shortages and lack of planning meant that unseasoned timber was often used.

During the late seventeenth and eighteenth centuries, as maritime war increased in Europe, rival nations vied with each other for vital supplies of oak timber and planking, as well as coniferous trees for masts and naval stores (chiefly pitch and tar for caulking timbers and weatherproofing rigging) from Scandinavia and the regions bordering the Baltic. Though pine

The Cutty Sark *was planked with teak and greenheart timber, bolted to wrought iron frames*

Manufacturing a timber-framed boat

timbers for masts were shipped from North America until the War of Independence (1775–83), enormous quantities of timber were taken from the Baltic between 1658 and 1814. However, British colonies in the tropics began to offer alternative supplies of wood and by the early 1700s bounties were being offered to merchants in an effort to encourage this trade.

By the beginning of the nineteenth century, iron and steel were being introduced into the hulls of merchant ships, but teak and other tropical hardwoods were highly valued. The famous British clipper the *Cutty Sark*, one of the fastest sailing ships of its time, built in 1869 to bring tea from China, was planked with teak and greenheart timbers which were bolted to wrought iron frames. Mahoganies and cedar from South and Central America and agba (*Gossweilerodendron balsamiferum*) from West Africa as well as other relatively light, decay-resistant timbers were also widely used around this time. By the end of the nineteenth century, however, most deep sea cargo ships had hulls, decks, spars and rigging made of steel, and the last major wooden passenger ship, the *Torrens*, set sail for Adelaide in 1903.

Today the framework of container ships and passenger liners and most large naval vessels is made of steel or aluminium. Smaller pleasure boats (generally below 18 metres in length) including various cruisers, yachts and sailing boats, are often made of glass reinforced plastic (GRP) which needs less maintenance and is often cheaper to buy and easier to position than timber. But wood, despite high labour costs, is still in use.

Modern preservative treatments, special epoxy resin glues and resilient paints have made marine plywood very durable, strong, and extremely versatile. It can be used to build a structure of any size — overcoming incidentally, many of the problems which hindered ship builders in the past. Wood veneers meanwhile can be moulded and laminated to almost any shape. The use of plywood and these veneers has reduced the amount of timber required for boat building and the degree of waste.

For racing boats in particular, marine plywood can offer a much better strength to weight ratio than some of its competitors. Most modern wooden boats are built using the 'shell first' method (once practised by the boat builders of ancient Rome), in which frames are laid inside the completed hull. For a small cruiser, for example, the frame may be built up of separate 'skins', individually contoured and shaped by heating on a mould, then glued in alternate directions for maximum strength. Woods used might include Honduran mahogany and West African obeche with the outermost ply sometimes laid horizontally to give the impression of traditional carvel planking.

The shell of the cruiser may also be made by a process of 'cold moulding', that is, by lying long, thin strips of wood diagonally over one another. The advantage of this method is that a

Balsa wood rafts on the Ecuadorian coast. This is a pre-Colombian design still used by fishermen today

Wooden boats being built in Greece

variety of softwood timbers can be used, such as Douglas fir or different pine species. Interior frames, positioned to strengthen the hull are also often made of laminates.

Whilst decking timbers might be made of larch, Scots pine or cedar, interior joinery woods once commonly teak, oak or mahogany, have been added to and replaced in some cases by cheaper African hardwoods such as utile, iroko, and opepe.

For lifeboats and fishing vessels, built for the roughest seas, wood was once pre-eminent. All new lifeboats are now made from glass reinforced plastic, but some fishing boats are still made from wood. In Britain the building of wooden trawlers is now only carried out in Banff in north-east Scotland. Whilst the frames of these boats are made from solid oak as in the past, the carvel style of construction has been adopted for their outer planking, that is assembling side planks (cut from Scottish larch) edge to edge, and caulking the seams.

Despite their impressive size and exclusive use of metals for construction, modern passenger liners such as the 'QE2' rely on wood extensively for interior joinery and decking. The decks and handrails of this grand ship are in fact all made of teak.

PLANTS IN THE RIGGING

Knowing the ropes!

From the earliest times different peoples have made rope and cords from locally available materials to help them build, sail, load and moor their sea and river-going craft.

Until the last 50 years or so, with the exception of wire cables developed at the start of the nineteenth century, plant fibres were the only practical source of cordage for the world's shipping, both on a large and small scale. The development of man-made materials and machinery to maximize production has greatly diminished the modern application of natural fibre ropes at sea.

Today's ocean-going ships and tankers as well as smaller pleasure boats of almost every kind, including speed boats, yachts and cruisers, rely chiefly on synthetic ropes made of nylon, polyester, polypropylene and polyethylene. The high-strength properties, immunity to rot, mildew and marine decay, and low water absorption of these ropes have all favoured their adoption.

However, several plant fibres of great historical importance are still being used by rope-makers in Britain for a variety of nautical applications: hemp from the stems of *Cannabis sativa*; Manila hemp from the leaves of *Musa textilis* a relative of the banana; sisal from the leaves of two *Agave* species, and jute and flax, stem fibres from *Corchorus* spp. and *Linum usitatissimum* respectively.

Of these, Manila hemp and sisal are now the most important for commercial use in shipping. Whilst neither is as strong as their equivalent made from synthetic fibres, their hardwearing and low stretch characteristics, plus their resistance to heat and sunlight have ensured their continued use: Manila, usually for capstan and winch operations and sisal for the mooring of small craft, lashing and cargo handling. Jute, hemp, and flax (and sometimes cotton) are still made into smaller sized ropes and cords which have a variety of general uses on board.

For many centuries the traditional fibres for ships' rope and rigging in Northern Europe were home-grown hemp. Tarred anchor cables were also often made from this resilient fibre until the early nineteenth century. Whilst steel wire came to fulfill many of the hemp rope's former functions, other fibres from the tropics were also being introduced. Coir ropes, for example, of very large sizes, made from coconut fibre were used for mooring because of their great elasticity and ability to float.

Many thousands of years ago the properties of several other plant fibres were utilized to great effect in Southern Europe and North Africa. The Egyptians twisted alfalfa grass, papyrus and date-palm fibre — as well as camel hair — to make ropes for an impressive range of barges, boats and warships. Some Egyptian galleys relied on ropes not just to set their sails but to give their vessels rigidity and to reinforce the sides. They also used flax, as did the Greeks and Romans, who added hemp and esparto grass to their range of raw materials.

Ropes used for mooring and rigging were traditionally made from plant materials including hemp and sisal

Musa textilis, Née.
1. Pistillate flower.　2. Staminate flower.　3. Fruit.　4. Section of fruit showing seeds.

ROPES

AGAVE

Agave sisalana and *A. fourcroydes*

Two species of desert-loving *Agave*, both native to Central America (*A. sisalana* and *A. fourcroydes*), produce very similar fibres both often known commercially as sisal. *A. sisalana*, however, is the source of 'true' sisal whilst *A. fourcroydes* produces henequen.

When Spanish conquistadores arrived in Central America in the early sixteenth century they found fibre from the plants in common use for making cordage and rough garments. Today the fibres are considered too coarse for clothing and their major application is rope and twine.

The fibres, which are extracted from the leaves, are classified as 'hard' since they are comprised of cells that are heavily impregnated with lignin (the compound that helps give wood its strength) and can be very laborious to remove by hand. For commercial production, leaves are cut from mature plants (in batches of between 40 and 80 leaves) and are next crushed between rollers. The resulting soft mush is then scraped to reveal the fibres, which are washed and hung to dry in the sun. A creamy white colour when properly washed, they can be dyed or used directly.

Though the fibres are not exceptionally durable, their length (often around 1 metre) and great strength have made them very important cordage materials, supplying approximately half of all the hard plant fibres in the world.

Whilst henequen is still grown primarily in its native Mexico, sisal is now being cultivated in large plantations in Brazil and East Africa.

MANILA HEMP

Musa textilis

Native to the Philippine Islands, *Musa textilis* is closely related to the banana. Its fibres (known both as Manila hemp and abaca) are extracted mainly from the edges of the leaf bases or petioles that form the 'stems' of these giant plants, by crushing and then scraping away the pulp. The fibres are certainly versatile and were used extensively in the past for making clothing in the Philippines, and later on in other parts of the world for making cigarette filters, tea-bags, dollar notes and salami wrappings!

Today most manila hemp is made into rope, its great virtue being a resistance to salt water. This has made it very suitable for marine work and enables it to compete with the most modern synthetic fibres.

Today most of the hemp used for rope making is imported from the Indian continent, while Manila hemp has come largely from the Philippines. Sisal meanwhile is imported from East Africa, Brazil and Mexico.

Sea and river travel are not the only forms of modern transport to make use of natural fibre ropes. The steel wire cables that raise and lower modern elevators need a fibre core for lubrication purposes and to help give them their vital flexibility. Though this may be made of synthetics, sisal is very often used — the choice depending on availability and customer preference. The fibre core acts essentially as an oil sump lubricating the cable and thereby reducing the frictional forces exerted on it as it travels to and fro over pulleys and other machinery.

DOCK CONSTRUCTION

The strong, durable timbers of a number of temperate and tropical hardwoods and some preservative-treated softwood trees continue to play an important part in general marine and fresh water construction work.

Dock and lock gates, piling, beams, and decking planks for wharves, jetties and piers as well as rubbing strakes and fenders which absorb impact and give protection still rely to a large extent on the special natural attributes of certain timbers.

Amongst the tropical hardwoods, greenheart (*Ocotea rodiaei*), opepe (*Nauclea diderichii*), iroko (*Chlorophora excelsa*) and ekki (*Lophira alata*) have been widely used. The great density of these timbers, their resistance to marine borers and to abrasion, and their availability in large sizes make them especially suitable for marine work.

Amongst the temperate hardwoods, oak still offers exceptional qualities. After preservative treatment, however, softwoods such as pine and fir species are also useful. The docks at Port Newark in New York are set on piles made of treated Douglas fir and protective fenders made of decay-resistant oak.

GREENHEART

Ocotea rodiaei

The unique properties of greenheart make it the ideal timber for dock and harbour work. It is one of the world's heaviest, hardest, strongest and most durable hardwoods and it is also resistant to marine borers.

Greenheart is often available in lengths of up to 17 metres and has a high commercial value. It is cut mainly from trees which grow in the rainforests of Guyana, though smaller quantities are exported from Suriname and Brazil.

Greenheart tree being felled in Brazil

PLANTS FOR ENGINES

The liquid fuelling our cars, indeed most internal combustion engines, and the oils that lubricate them are at least in part derived from plants. The petrol and oil used by millions of vehicles each day is the result of the compression over millions of years of minute sea-dwelling animals and plants that once lived on earth, and the mud and sediments that covered them as they died. Subjected to intense pressure and heat, this 'organic soup' underwent chemical and physical changes to end up as droplets of oil. The migration of these droplets through porous rocks and fissures led to the eventual formation of large underground reservoirs contained by layers of impervious rock, which are the source of most of our modern fuel oils, lubricants and petrol today. The earth's crude oil has also given us, of course, power, heat and light for industrial and domestic use and the petrochemicals from which thousands of items that we take for granted daily — from pharmaceuticals to plastic tea spoons — are made.

Whilst it is hard to equate the tiny sea plants that existed in pre-history with the petrol and oil in our cars today, some much more familiar plants are helping us to extract crude oil from the earth. As the hollow, rotating drill cuts down through impacted rock towards the oil deposit, a specially prepared mud or sand is pumped down the drill pipe under pressure to both lubricate and cool the bit and to force up to the surface, round the outside of the pipe, the rock debris dislodged by the drill. This mud is prepared with the addition of plant-derived materials. Potato starch, palm and coconut oil derivatives and a number of plant gums, including gum ghatti exuded from *Anogeissus latifolia* (a tree native to the dry deciduous forests of India and Sri Lanka) and guar gum, from the seeds of *Cyamopsis tetragonolobus*, make the mud viscous and relatively fluid by absorbing water. This enables it to mix with the rock drilled from the base hole and push it up to the surface.

Drilling for oil

DYNAMITE FROM PLANTS

The dynamite used by geophysicists in their preliminary searches for oil by means of seismic surveys is derived in part from ancient plants.

The explosive base of dynamite, liquid nitroglycerine, is by itself very unstable and liable to explode easily. To stabilize it and make it safe to handle, it is mixed with kieselguhr, a fine, earthy material mined from various sites in Europe, North America and Australia.

Kieselguhr is in fact composed of the fossilized silica coats of minute brown algae that lived in vast numbers in fresh and marine waters millions of years ago. Under the right conditions they formed thick deposits and beneath great pressure from accumulated mud and rock formed this highly useful porous 'earth'.

Lubricants from plants

Whilst lubricating oils for the majority of motor vehicles in use today are made from refined petroleum, engines that must work at very high temperatures to power our fastest forms of travel rely to a large extent on the oils produced by living plants.

Racing car and aeroplane engines — including Concorde's — now use lubricating fluids made in part from coconut and/or castor bean oils! Ordinary mineral oils cannot survive the intense heat generated by today's very powerful engines and as these were developed, it was found that special oils needed to be made to match their new capacity.

Before the development of the gas turbine engine, aviation piston engines had in fact used castor oil based lubricants. Their formulation, however, meant that they were not suitable for high performance engines and the search began for an alternative which had the properties of natural oils whilst eliminating their tendency to gel at low temperatures and to form gums and lacquers.

Mineral oils — which were used to lubricate most early jet aircraft engines — seemed to solve the problem for a while, but their use was unsatisfactory as it led to problems of deposition and degradation, frequent oil changes, increased oil consumption and higher mechanical failure rates. It was the simultaneous development of the new gas turbine engine and its 'polyol' ester-based lubricants, made in part from fractionated plant oils, that provided the answer — improving thermal stability and general engine performance, and lowering maintenance costs.

From about 1960 aircraft powered by gas turbine engines progressively replaced their post-war piston-engined counterparts in airline operation. Today these engines power most military and civil aircraft and are lubricated by the new 'synthetic' aviation oils, comprising mixtures of organic 'polyol' esters and various anti-oxidants, anti-wear additives, corrosion inhibitors, dispersants and anti-foam agents.

The esters are made by reacting organic acids with alcohols. The chief sources of the acids currently used are vegetable oils and animal fats whilst the alcohols are derived from petrochemicals. Of the vegetable oils, coconut — pressed from the white flesh of the nut — and castor — pressed from the bean or seed of the tree — are most widely used.

For racing cars, model aeroplane engines and speedway bikes that run on methanol, castor oil is used in much larger proportions. In fact it makes up 95 per cent of Castrol M, and around 50 per cent of the Castrol R brands.

Some car engine oils contain natural plant ingredients

PLANTS FOR FUTURE FUEL

It is not just lead that can be taken out of motor vehicle engines to make them environmentally safer. We could dispense with petrol — that is fossil fuels — altogether and rely much more directly on alternatives from plants!

In Brazil one quarter of the cars now sold run on pure ethanol, the alcohol fuel ('gasohol') made by the controlled fermentation of plant matter. Sugar cane stems and leaves have been chiefly used since the beginning of the country's National Alcohol Programme in 1975, but cassava now supplies a small proportion of the raw material. One drawback is that large tracts of Amazon forest have been destroyed to grow much of the sugar cane needed to produce the fuel, and the main by-product, a very rich organic sludge, has been dumped in rivers causing serious pollution. Use of cassava should help overcome the first of these problems since it will grow on dry and nutrient-poor soil (large areas of which already exist due to past deforestation) and this should avoid the need for more forest destruction.

At least 46 countries are currently operating 'biomass' production programmes in an effort to achieve at least a degree of self-sufficiency in energy production, for a number of purposes. A variety of different plant materials and crop residues are in use, including maize, sweet sorghum, sweet potatoes and sugar beets. In China, where efficient farming methods ensure that very little is wasted, banana leaves as well as sugar cane are used to produce 'biogas' and such prolific and tenacious plants as water hyacinth and Napier grass are other contenders.

Away from dry land, seaweeds offer great potential for alcohol production. Giant kelp (*Macrocystis pyrifera*), some individual lengths of which are said to reach 200 metres, and which grows in extensive beds, has already been used in experimental 'ocean farms' off the Californian coast. It has been estimated that one square mile of ocean farm could produce enough energy to support the yearly needs of 300 people — supplying not just fuel, but fertilizer and extensive stocks of fish.

Back in the Amazon rainforest, a number of individual trees have been found to produce oils that can be used directly as fuel. The liquid balsam stored within the copaiba tree (*Copaifera* spp.) was discovered in 1979 to be almost identical to diesel fuel, and capable of powering a diesel engine simply by being poured into its fuel tank! The traditional use of this precious liquid was for years as a raw material for the perfumery and pharmaceutical industries of North America and Europe. Whilst the trees are now being grown in experimental plantations in Japan, the great potential of copaiba, which can produce four gallons of 'fuel' in 2 hours, is as yet unrealized.

As its name suggests, the petroleum nut tree (*Pittosporum resiniferum*) from the Philippines also produces a high octane oil that can be used directly as a fuel. During World War II, the Japanese used it to power their tanks, though tribal people of the Philippines have long appreciated it as a fuel for lamps. The oil is pressed from the nuts or seeds of the tree. Just six trees are said to be able to produce some 320 litres of oil in one year.

Much less exotic plants could also help solve our energy problems. As Rudolph Diesel wrote in 1911: 'The diesel engine can be fed with vegetable oils and [this] would help considerably in the development of agriculture of the countries which will use it'. A 3 to 1 mixture of sunflower oil and diesel is in fact very similar in efficiency to ordinary diesel, and many other oils such as those pressed from olives, peanuts, soyabeans and palm-fruits, as well as sunflower, rape and sesame seeds and eucalyptus leaves could be used.

*The copaiba tree (*Copaifera sp.*) stores a liquid balsam, which is almost identical to diesel and has great potential as a future fuel*

Recreation — plants that entertain us

Paintings from ancient Egypt, on papyrus

Whether doing a crossword or reading a book, playing cricket or a double bass, painting a picture or taking one, plants are there again, giving us the raw materials, providing countless means of entertainment and communication.

Plants are essential for a tremendous range of recreational activities, especially in the fields of sport and music, where different woods in particular have played a vital role. Often the fact that we are using substances derived from plants is not at all obvious. Almost all photographic film, for example, is made from a base of plant cellulose, yet this is the last thing we are likely to think of as we load the camera. Perhaps the most commonly used material of all, for recreation of any kind, and of course for communication, is paper. This is also made from cellulose. The artist, book-worm and globe-trotter, armed with paper currency or travellers cheques, would all be lost without it. Even the most sophisticated word processor needs something to print its binary message on. In the main, they all use trees.

PAPER — PLANTS FOR WORDS AND PICTURES

In 1987 we used over 214 million tonnes of paper and board world-wide, made from 156 million tonnes of paper pulp. Most of this pulp was made from de-barked hard and softwood tree trunks grown specially for the purpose.

Wood has not always been the source of paper though, and as basic pulp-making technology has changed so have the raw materials. At least 5500 years ago the Egyptians were using thin strips of papyrus peeled from the stems of this reed-like plant to write on. The strips were laid side by side and over one another at right angles and after moistening were pressed flat and left to dry. The dried material was hammered to make it more compact and rubbed with a stone or bone to produce a smooth surface. To make a scroll, several of these flat sheets would be stuck together. Though the use of papyrus was certainly very significant, having left us not just with the word 'paper' but innumerable writings from Egypt and the Middle East, and most of the works of the great Greek and Roman scholars, paper as we know it now had its origins in a significantly different use of plants.

Paper history: from wasp's nest to glossy magazine!

By at least 200 BC the Chinese had begun to make fine sheets of paper, influenced, it is said, by the nest-building activities of wasps. The brittle paper made by wasps is basically the result of redistributing the cellulose fibres contained in wood or woody matter by breaking down the sticky substances that hold them together and leaving the fibrous mass to dry. Though paper-making is now a highly sophisticated operation relying largely on chemicals to produce the basic pulp and a variety of additives to give the customary finish to the sheet, these underlying

Papyrus growing by the edges of a lake

principles have hardly changed. The Chinese began their operations using the inner bark of the paper mulberry (*Broussonetia papyrifera*), hemp waste and that most versatile giant grass, bamboo. Expanding on the work of the Chinese, the Arabs developed the practice that was to last until the nineteenth century of using hemp, cotton and flax in the form of rags as their chief raw materials. As they came to dominate different lands, their technology of paper-making went with them, replacing the use of papyrus in Egypt and the Mediterranean countries. Indeed, it was the Arab invasion of Spain in the eleventh century and the dissemination of their skills and learning that brought paper-making to Europe.

The linen, hemp and cotton rags used were dampened until they began to rot, then cut into strips and softened with more water in large vats. Huge wooden beaters — often worked by water mills — broke the rags down, separating the fibres and breaking their cell walls into tiny fibrils. The fibres were then mixed with water again and various proportions of glues and resins to make the paper pulp. A rectangular frame with a fine sieve at the bottom was plunged into the mixture and pulled out by hand causing the fibrillated fibres to form a sheet as the water drained away. Several thin layers made in this way and pressed after removal from the frame would go to form a finished sheet of paper.

It was a further observation of the work of wasps, this time by the Frenchman René de Réaumur in the early eighteenth century, that helped to turn the 'industrial eye' to wood for paper making. However, the commercial substitution of logs for rags, which were becoming increasingly scarce as the demand for paper rose, did not take place until the early nineteenth century. This major change was to be preceded by another of the greatest importance — the mechanization of the paper-making process. In 1803 the Fourdrinier brothers successfully developed and patented a machine, invented earlier by one Nicholas-Louis Robert, that could deliver a continuous sheet of paper to a pair of rollers. The paper-making process was now revolutionized in terms of the size and quantity of sheets that could be made, and this development helped considerably to meet demand. But the paper mills were plagued by shortages of rags and a new material needed to be found.

Paper is still handmade by some craftsmen, from a variety of different plants. Here the plant being used is New Zealand flax, Phormium tenax. The leaves are cut and pulped, then a rectangular sieve is pulled through the pulped mixture to form a sheet. This is then pressed to get rid of the water.

Different types of paper are used for everything from newspapers to art materials

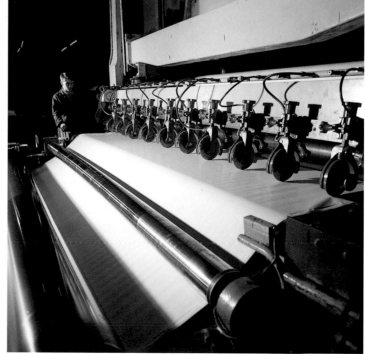

Machines can produce vast quantities of paper in continuous rolls

MEXICAN BARK PAPER

In various parts of Mexico, papel amate — bark paper — was being made over 1400 years ago by many indigenous groups. The Aztecs, Toltecs and Mixtecs all produced fine paper sheets and large quantities were sent by way of annual tribute to Montezuma II, the Aztec ruler.

The sheets were used for making special books (codices) in which important mythical and historical events were recorded, for adorning temples and altars and for use in religious ceremonies.

Bark paper is still being made today by the Otomi people of San Pablito in the state of Puebla, but theirs is the only paper-making centre left in Mexico. Until the recent popularization and commercialization of their craft, the Otomi used the bark of various fig tree species, known locally as amate — chiefly *Ficus tecolutensis* — and mulberry, (known as moral) — mainly *Morus celtidifolia*. The over-exploitation of these trees however, plus changes in land use from forestry to commercial agriculture necessitated the substitution of other kinds of bark. Today jonote colorado bark (*Trema micrantha*) is chiefly used, from trees grown in the state of Veracruz.

To make the paper, the fibrous inner phloem fibres are separated from the outer bark in strips and boiled for several hours in water containing lime. This softens the fibres and makes them separate more easily. After rinsing, the strips are arranged in a grid-like pattern on a smooth board and then beaten with a special flattened stone, the muinto, until the fibres mesh together. Once they are sufficiently intermeshed, the newly formed sheet (still on the board) is left to dry in the sun.

As the object of a flourishing tourist trade and instantly recognizable today by their often fluorescent illustrations, bark paper pictures are to be found not just in Mexico but adorning countless homes around the world.

About 2500 people in San Pablito currently make all or part of their living by producing papel amate.

Making Mexican bark paper

With the German development in 1840 of a means of reducing logs to pulp by holding them against a revolving grindstone beneath a jet of water, the earliest mechanized technique of making paper pulp from wood was born. Shortly after this, chemicals were introduced to do the job (much more effectively for certain kinds of pulp), and both these processes, the chemical and mechanical, are used today.

The main advantage of chemicals was that they dissolved the lignin and other softer components of the wood without damaging the cellulose fibres as mechanical grinding had done, leaving a pulp of far superior strength. The most modern processing plants produce enormous quantities of chemical pulp chiefly using the sulphate or 'kraft' method, also invented in Germany. In this process wood chips from almost any tree species can be reduced to pulp by first steaming and then saturating them with hot sodium hydroxide or sodium sulphide at very high temperatures. After washing and 'flash' drying, the fluffy crumb-like particles which emerge are bleached before conversion into all kinds of writing, printing and drawing papers.

Mechanical means of reducing logs to pulp, by grinding them on giant wheels or pulp stones, are used for the production of coarser papers such as newsprint. Since this pulp will contain almost all the wood's components — such as lignin, hemi-celluloses and resins — and not just its fibres, it is not suitable for high quality paper. Paper made from this pulp tends to turn yellow when exposed to heat and light for any length of time — as can be seen with old newspapers and paperback books.

To convert it into paper, the chemical or mechanical pulp, or a mixture of the two depending on the type of sheet required, is first fed onto a fast-moving screen of wire mesh on which it drains, leaving the fibres interlocked. It then passes through a series of heavy rollers and heated cylinders which remove most of the remaining moisture and press the paper absolutely flat, to produce a continuously emerging roll.

Trees for pulp

Naturally enough, the modern production of wood pulp and paper has tended to be concentrated in highly industrialized, well-forested countries. For the last twenty years Canada, the USA and Scandinavia have produced over 80 per cent of the world's pulp — chiefly using softwoods — but as different countries have begun to exploit their natural resources the sources of supply have also changed. Whilst the range of trees used for pulp-making has expanded, the huge demand for paper world-wide, not just for books and magazines, but industrial packaging and household goods, has led to the search for faster growing 'super trees' which can be genetically manipulated to give a high yield of fibre under intensive cultivation.

Over half of the sulphate paper pulp imported by Britain today is made from eucalyptus trees of two main species, *Eucalyptus grandis* and *E. globulus*, grown in Brazil, Spain and Portugal. In recent years Brazil has planted huge areas in the Amazon with cloned eucalyptus seedlings, specially selected and engineered so that each tree will be genetically identical and ready for harvest after about seven years.

Apart from eucalyptus, the hardwood pulp will include a proportion of temperate species grown in northern Europe, such as birch, beech, chestnut, poplar and aspen — the latter now becoming a popular choice for plantation cultivation in Canada — and a smaller percentage of mixed tropical hardwoods, most likely to be coming currently from South East Asia. Softwood pulp, on the other hand, is made from various spruce and pine species grown chiefly in northern Europe and North America.

*Monterey pine (*Pinus radiata*) is a fast-growing softwood used for paper pulp*

Overall, the UK's white chemical paper pulp is comprised of around 70 per cent hardwoods and 30 per cent softwoods. What is bought depends to a large extent on price and availability and the position of the dollar. In the USA the percentages of hardwood to softwood used are a rough reversal of those used in Britain, with most pulp coming from native softwood trees, many of which are known collectively as 'southern pine'. These trees, chiefly *Pinus taeda* (loblolly pine); *P. palustris* (longleaf pine); *P. echinata* (shortleaf pine) and *P. elliotti* (slash pine) — the last three sometimes exported under the name of pitch pine — have been genetically 'improved' to have a higher wood density, to be highly resistant to disease and insects, and to be harvestable five years earlier than ordinary pines.

Similarly 'improved' *Pinus patula*, *P. taeda* and *P. elliotti* trees are being grown for pulp production on a massive scale in southern Africa. In Swaziland some 70 million pine trees now cover 65 000 hectares of what was once open grassland. On average the trees are harvested at 15 to 20 years of age, having reached a height of around 25 metres. Large stands are clear felled at a time and some 4.5 million seedlings are planted out each year.

Plantations of another very useful fast-growing softwood — *Pinus radiata* (Monterey pine) — have now been established in Chile, New Zealand and Australia.

Different plants for different paper

All plant fibres have their own special characteristics, and these contribute directly to the quality and type of paper that can be made from them. Softwood fibres are generally longer than hardwood fibres, and since they intermesh over a greater area, they make a stronger paper. Whilst the shorter hardwood fibres make a weaker pulp — often forming, for example, the majority of the pulp for household and toilet tissue — they give a better surface for printing on. A typical sheet of writing paper made in Britain could well comprise 55 per cent hardwood pulp (a large proportion being eucalyptus), around 20 per cent softwood pulp, and 15 per cent filler, chiefly chalk and special clays.

It is not just trees, however, that make the basis of the paper pulp in use today. Speciality papers of very high quality often comprise a proportion of esparto grass fibres (*Stipa tenacissima*). Though they are very short and have the smallest diameter of all the common paper-making fibres, they give a bulk and opacity to paper, plus a closeness of texture and smoothness of surface which are unique. Paper made from esparto also water-marks very clearly and expands less than other papers when wetted. The esparto grass used in Britain today is imported from North Africa and Spain.

For paper that must be very strong yet flexible, such as bank note paper and some legal documents, mixtures of hemp, flax and cotton fibres — all of which are longer than those of wood and need far less drastic processing — are used, but mostly, as in the past, in the form of rags. Almost the whole of the world's production of raw cotton fibre — the purest form of natural cellulose — and linen is used by textile and yarn manufacturers so paper makers are obliged to use them in this condition. Some cotton linters, the shorter seed fibres, and a certain amount of flax waste or 'tow' too short for spinning, are also incorporated.

In other parts of the world, locally available plants will often determine what different nations make their paper from. At least one pulp mill in Denmark processes straw from various temperate cereal crops, whilst China as a whole uses a large amount of rice straw. Along with India, Mexico and Brazil, China also makes extensive use of bagasse, the waste from sugar cane, but another plant accounts for far greater quantities of paper overall — bamboo.

Though the uses of bamboo are legion, more ends up as paper than as anything else! Currently between 7 and 8 million tonnes of raw bamboo are pulped world-wide each year. China, Thailand, Taiwan, Indonesia, Burma and Bangladesh all convert huge quantities, but the single biggest producer of bamboo for pulp today is India, where some 35 factories turn about 1.75 million tonnes into this basic raw commodity each year. The fibres have a high cellulose content and since they are more slender than wood fibres can produce a smooth, flexible paper. The only disadvantage in using bamboo is its high processing cost since the pulp has more impurities than wood pulp.

Though careless over-exploitation in the past has caused problems in some regions, the various bamboo species suitable for pulp can and do grow so prolifically over so wide an area that the future of the plants continues to look promising, not only for those Asian nations already cultivating them, but for paper makers in other areas where timber is not easy to obtain.

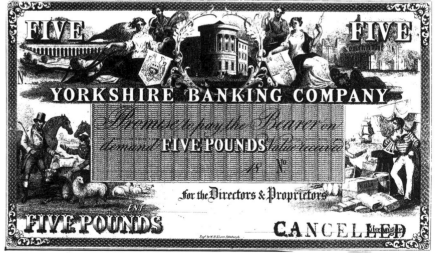

Five pound bank note of the Yorkshire Banking Company, established in 1843.
Bank note paper is manufactured from cotton or linen rags

MORE PLANTS FOR PAPER

Before we come to lay our hands on finished books or magazines and the paper money that we use to buy them, plants are used again in two important ways: firstly to make sizing agents which improve the paper and prevent it from becoming too absorbent, and secondly as components of printing inks.

Sizing agents

Though synthetic sizes have now been developed which are effective in the strongly alkaline paper pulps, several plant-derived size materials are still in use — chiefly natural starches, gums and resins. The most widely used include potato starch, guar gum, methyl cellulose, alginates and rosin. When mixed into the pulp the selected size is usually just one of various additives such as dyes, brighteners and fillers (chiefly china clay and chalk) which will give the desired characteristics of colour, smoothness and opacity to the paper. In this form, or as an external coating, the size helps to determine the moisture resistance of the paper to be printed. If it is left out, too much ink will penetrate the paper and leave it smeared.

The various plant sizes all have different attributes which will naturally give differing results. Methyl cellulose for example has excellent oil and grease resistance properties and also helps to give an even finish. Alginates too improve the paper's surface and give added smoothness as well as better ink acceptance, whilst potato starch, available in different formulations as an additive for pulp or as a surface size generally improves the paper's moisture retention.

The main use of guar gum in paper-processing is as an additive to pulp. The chemical pulping process removes not just the lignin from the fibres but a large proportion of the hemi-celluloses present in the wood. Guar gum, and to a lesser extent locust bean gum, replace or supplement the effects of natural hemi-celluloses in bonding the fibres together and distributing them more evenly. This gives a better sheet formation and improves folding, bursting and tensile strengths. The gums, used traditionally in a variety of paper pulps, also make the paper less prone to curling at the edges.

Pine trees also help achieve the qualities of many of the paper products that we use. Rosin — a natural resin — is obtained commercially today from a variety of pines grown mainly in the USA, China, Russia and Europe. Added as a size to the paper pulp, rosin helps improve the finished surface of the sheet and enables it to be printed on without the ink smearing or 'feathering'.

Left, a four-colour printing press and below, a newspaper being printed with black ink

GUM ARABIC

Acacia senegal

More than 500 species of *Acacia* tree are distributed throughout the tropics and subtropics, but most of the world's gum arabic is obtained from only one of these, *A. senegal*. More than 90 per cent of the gum we use is collected by local people from wild trees, which grow in the sub-Sahara and Sahel regions of Africa. The bark of the spindly, spiny trees is cut with a knife and a 'tear' of amber gum forms slowly at the scar. After harvesting, the gum is cleaned and sorted into various grades for export. Processing — into milled or spray-dried powder — is usually done subsequently by the importer.

Gum arabic is non-toxic, odourless, almost colourless, tasteless and water-soluble, and as such has found very wide use in a number of important industries. It is used to thicken many convenience foods, as well as pharmaceuticals and cosmetics, and may appear in textile sizes, water-colour paints and printing inks.

Gum arabic was used until relatively recently to make the glue that coats the back of postage stamps. It is still added to some commercial beers to help prevent the foam that crowns the glass from subsiding!

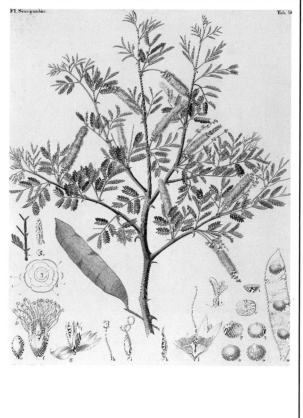

Plants for printing inks

Many printing inks rely on plant materials to give them their distinctive properties. Linseed and tung oils, rosin, gum arabic and fractionated palm and coconut oils are the most likely candidates.

Classed as 'industrial' oils, linseed and tung oils are especially useful to ink manufacturers since they will dry naturally, an attribute which has made them suitable for use in various paints and varnishes too. Tung oil, which comes from the seeds of several species of deciduous *Aleurites* trees native to China, was an original ingredient of 'India ink' still widely used by China's calligraphers.

The printing industry also currently consumes large quantities of rosin and rosin derivatives for many different kinds of ink. These additives provide adhesion and give gloss and hardness to the product as well as helping it flow freely.

Gum arabic is added as a binder (fixing the solid pigment particles to the surface to be printed) and suspending agent, controlling the consistency of the ink and the settling of the other ingredients. Early industrial inks were simply dispersions of lamp black in water, but gum arabic was soon added to improve their texture. The Egyptians were, however, the first to appreciate the qualities of this very valuable plant exudate (produced only when the tree's tissues are in an unhealthy state), using it to suspend the pigments in their coloured inks.

Today an important function of the gum in some speciality and typographical inks (made from oil-in-water emulsions) is to stop them from penetrating the paper too deeply. The lithographic printing process was developed in 1768 with the aid of gum arabic and it is still vital for this method of printing today. It acts as a desensitizing agent by making the image to be printed receptive to oil, thereby distinguishing it from the non-printing area which is oil repellant.

An artists' tools — many incorporate materials from plants

PLANTS FOR BUDDING ARTISTS

Paints

Artists and illustrators everywhere and all of us who like to leave our own impressions on paper using paints or crayons, have been helped by a variety of different plants to do so.

Linseed oil, already mentioned for its use in printing inks, is still an important ingredient of artists' oil paints as are poppyseed and sunflower seed oils (which are lighter in colour), acting as the basic medium or binder which fixes the pigments to the canvas or other surface to be painted.

Though oil varnishes made by dissolving resins in a drying oil were used in antiquity in Italy and Greece, Flemish artists are credited with developing the forerunners of the oil paints we use today by combining plant oils with powdered pigments. In the fifteenth century the Van Eycks are said to have perfected a technique of applying pigments in the form of thin translucent glazes made from a base of walnut or linseed oil. The effect was sensational since it produced a feeling of texture and dimension that had not been seen before.

The luminous depth characteristic of many oil paintings has been achieved with the help of two other plant materials: copals and dammars. These are resins, either fossilized or recently tapped, that are sometimes used to varnish oil paintings. Applied in liquid form, they effectively 'hold' the paint when dry and give it a special lustre and transparency. Copal is produced by species of *Copaifera* and the related genus *Hymenaea*, and by trees belonging to the genus *Agathis*. To obtain a fresh supply, the bark of the trees is slashed and left to ooze the resin until a mass has accumulated over the cut.

Dammars are obtained in the same way but primarily from species of *Shorea*, a genus of the Dipterocarpaceae family. Both copals and dammars have been utilized historically in many paints and lacquers, but their use is now practically confined to oil paint varnishes.

As an oil paint thinner and solvent for brush cleaning, turpentine from pine trees has long been indispensable, but in recent years the development of cheaper, petroleum-based solvents ('white spirits') has brought a sharp decline in this traditional use. Some turpentine is still available for artists' use though, and in the USA much wood turpentine is derived from longleaf pine (*Pinus palustris*). Today most turpentine produced is further processed into its

MADDER

Rubia tinctorum

Rubia tinctorum is one of some 38 species in the genus *Rubia* and is the largest of the bedstraws. Native to Eurasia, it has trailing, angular stems with narrow, evergreen leaves that are dark and shiny, arranged around the stem in whorls. Small yellow flowers are followed by blackish berry-like fruits. The thick, fleshy root — which produces madder dye — has a dark rind with a ruddy-coloured inner part. Wild madder, *R. peregrina*, grows in France and south-west Britain, but produces a weaker red dye than *R. tinctorum*.

Wild madder (Rubia peregrina)

Rubia tinctorum

various chemical constituents, and in this fractionated form is used to make a number of organic chemicals, chiefly for fragrance, flavouring and vitamin manufacture.

Whilst water-colour paints do not require plant oils or turpentine for their manufacture, they do use other plant materials. Gum arabic is an important ingredient of most smudge-proof water-colours, since in acting primarily as a binder, it keeps the particles of solid pigment fixed to the paper. As it absorbs water to form a paste, gum arabic also holds the pigments in suspension so that they can be evenly applied. Dextrins made from maize and potato starch are also used to bind and viscosify the paints' ingredients. Like gum arabic , a traditional water-colour medium — as yet unsurpassed in quality by modern chemical equivalents — these water-soluble materials are especially valuable because of their non-toxicity.

The pigments used in water-colours, oil paints and most other artists' materials, whether coloured pencils, chalks or wax crayons, are almost all derived today from chemical compounds containing different metals such as cobalt, iron and lead, and from petroleum. Two exceptions do remain, however: madder, for distinctive pinks and reds, and gamboge for yellow. Madder is extracted from the crushed roots of the herbaceous perennial *Rubia tinctorum* whilst gamboge is prepared from a natural exudate of the South East Asian tree *Garcinia hanburyi*.

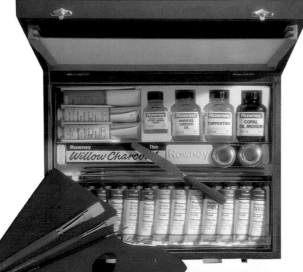

Oil paints have linseed, sunflower seed or poppy seed oil as the binder fixing the pigments to the canvas

Monks on monastery balcony, Burma, wearing robes dyed with gamboge

The madder grown commercially today comes chiefly from southern France and the Middle East. The fleshy roots are dried and their outer bark removed, then the inner portion is dried again and milled to a very fine powder. Both bark and inner root produce the colouring matter — which is extracted from the powder by fermentation and hydrolysis with dilute sulphuric acid — but the bark gives an inferior colour. Madder Lake and Rose Madder for artists' pigments are prepared from the extract by adding alum and mixing this with an alkali to form a precipitate. Madder plants were once widely cultivated throughout Europe and the Middle East and were being used by peoples of the Indus valley at least as early as 3000 BC. Cloth remnants and wall paintings found in ancient tombs also prove its long history of use in Egypt.

The plants are said to have been introduced to Italy and France at the time of the Crusades and to Holland in the sixteenth century by the Moors. Though they had become important as a source of dye for cloth and for medicinal purposes, it was not until the seventeenth and eighteenth centuries that pure madder pigments came to be adopted widely by European painters.

Natural madder contains the colouring matter alizarin, together with the closely allied purpurin and certain other pigments in small quantities. After the invention of synthetic alizarin from chemicals in 1868 the plant was largely ignored as a source of dye and artists' paint, but the preference of traditionalists for this colour and the fact that it is one of the most stable of the natural organic colouring materials has extended the use of madder as an economic plant.

Gamboge has been used as a pigment for centuries in the Far East. It was the traditional dye for the silk robes of Buddhist monks and priests and the unique brilliance of its colour made it the prime ingredient of certain yellow paints, as well as special varnishes and inks.

The name gamboge is a corruption of Cambodia (now Kampuchea), one of the chief countries from which the raw material was exported. The Dutch introduced gamboge to Europe during the mid-seventeenth century and it was subsequently used by early Flemish oil painters.

The pigment is contained within the resin produced by various species of tropical *Garcinia* trees — mainly *G. morella* and *G. hanburyi*, distributed throughout Thailand, Burma, Kampuchea, the East Indies, Sri Lanka and southern India. Only *G. hanburyi* however, has been exploited on a commercial scale. Once the trees have reached at least ten years old, a spiral cut is made in the bark during the rainy season and the resin which is then exuded is traditionally collected in bamboo lengths placed at the base of the cut. The resin is allowed to harden — a process which takes about a month — after which the bamboo containers are heated and the gamboge taken out in the form of cylindrical sticks. Alternatively the gamboge is collected simply in the form of 'tears' or fragments which form in the region of the incision. Though the raw resin is a reddish or brownish orange in colour, it turns brilliant yellow when powdered and gives a deep yellow emulsion when mixed with water. Whilst it fades rapidly in sunlight as a water-colour, it is fairly permanent when ground in oil.

Apart from its modern use in the best water colour paints, it is the principal colouring ingredient in certain golden spirit varnishes, and in Thailand a beautiful golden yellow ink for writing on black paper.

Garcinia hanburyi; *its resin contains gamboge pigment*

Pencils and wax crayons

Wax crayons and the cores of coloured pencils use plant-derived materials chiefly to bind all their ingredients together. Methyl and ethyl cellulose made from wood pulp or cotton linters may make up between 15 and 20 per cent of a crayon core, helping to produce an even flow of colour onto the paper, while various plant waxes are also likely to be present. Carnauba and candelilla wax, as well as Japan wax, a now very expensive material prepared from the berries of several species of sumac tree (*Rhus* spp.) cultivated in Japan and China, are mostly used as well as small quantities of natural resins.

The outside of a coloured pencil is usually coated with lacquers made in part from plant materials. Nitro cellulose — made from a base of wood pulp or cotton linters treated with acids — is the most common ingredient, though dammar resin and shellac, an encrustation made by *Tachardia lacca* insects chiefly on *Butea frondosa* trees, are also used.

As for the supporting body of the pencil itself, wood, of course, is the traditional material. One wood in particular became so much associated with pencil production in the past that it is still commonly referred to as the pencil cedar. It is a juniper species native to North America, *Juniperus virginiana*. Its wood once dry does not distort or splinter, and its straight grain and softness allowed it to be easily whittled or sharpened to a point. This raw material was used extensively for many years — indeed so much so that by the end of World War I it had become extremely scarce. In the 1920s, however, another conifer with similar suitability for pencil

INCENSE CEDAR

Libocedrus decurrens

This cedar, often very slim with dense, bright green foliage, but developing an open crown with level branches where its growth is unimpeded, now supplies around three-quarters of the world's wood-cased pencils. Harvested annually, the trees are grown commercially in forests of the Oregon Cascades and the Sierra Nevada ranges of California, where they are native. Their soft, light wood is easily sharpened without splintering, and is finely textured, making it perfect for pencil production.

*Most pencils are manufactured from incense cedar wood
Although most erasers are now made from
PVC, some are still manufactured from rubber
latex tapped from* Hevea brasiliensis *(right)*

manufacture was selected to replace it and is now the main supplier of wood for pencils: *Libocedrus decurrens*, the incense cedar, also native to North America.

The cedar logs are first cut into squares which will be seasoned for a year before the pencil-making process can begin. Each square is then sawn into blocks approximately 185 mm long, and these are sawn into the slats from which the pencil body will be made. The slats are now graded and treated under pressure with paraffin wax and a stain. Whilst the stain will give the finished pencil body a uniform colour, the wax lubricates it during manufacture. Grooves cut along the length of the slat hold the pencil 'leads' in place, and a sandwich made of two slats stuck together forms a solid block from which the pencil lengths are cut.

Over 70 per cent of the world's pencils are currently made from incense cedar grown chiefly in the forests of the Oregon Cascades and the Californian Sierra Nevada mountain ranges. Its straight grain, uniform texture and relative softness have made it very suitable for precision sawing, staining and waxing, and for the pencil user, for sharpening into a satisfactory point.

Erasers

When it comes to erasing pencil marks, we have traditionally turned once more to plants — specifically to that most important economic product made by several of them — rubber.

It was Dr Joseph Priestly's observation in 1770 of 'a substance excellently adapted to the purpose of wiping from paper the marks of a black lead pencil' that lead him to coin the name 'rubber' for this revolutionary substance.

At this time, not long after it had been brought to Europe by Charles Marie de la Condamine, but before its full potential had been realized (through vulcanization), lumps of raw rubber were being sold in London for use by artists and all pencil users. Whilst this practice continued long after other uses for natural latex had been found, the development of new synthetic rubbers from coal tar and petrochemicals (from the 1930s onwards) brought a change in the components of the household eraser.

Today most are made from PVC (polyvinyl chloride) but a proportion of those used by artists are still made from the rubber latex tapped from *Hevea brasiliensis* trees.

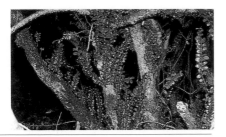

*Box (*Buxus sempervirens*) was once the chief material used for making drawing instruments such as rulers*

Drawing instruments

Like rubbers, rulers are most likely to be made today of man-made substances — generally transparent or opaque plastics — but for many years boxwood was the chief material used. The box *Buxus sempervirens*, native to the chalky soils of southern England, is often clipped to form an ornamental bush or dense evergreen hedge, but can reach a height of over 10 metres as a fully grown tree. It produces a hard, even-textured yellow wood, so heavy that it will sink in water when green. Its great stability, and the fact that it does not warp, made it very suitable for the manufacture of mathematical instruments such as rulers and set squares, which needed to be strong and which could be precisely marked.

In the eighteenth century box became especially popular for the blocks from which artists made their wood engravings. Other uses included the manufacture of chess pieces. The great demand for box though has led to the best and largest trees being felled, leaving only the less vigorous specimens in their natural habitat.

Two tropical timbers unrelated to *Buxus sempervirens* but with similar properties and available in larger sizes have been used more recently for mathematical instruments, including T-squares and rulers: ramin (*Gonystylus bancanus*) from Malaysia, Indonesia and the Philippines, and maracaibo boxwood (*Casearia praecox*), exported from Colombia and Venezuela.

A GREEN LIGHT FOR PHOTOGRAPHY — THE CAMERA AND PLANTS

The second most popular leisure activity in the UK — beaten only by gardening — depends almost entirely on plants.

Practically every roll of photographic film is made from a base of cellulose acetate, manufactured as we have already seen by reacting the cellulose from a range of soft and hardwood trees and cotton linters with acetic anhydride and other chemicals. To make film, the syrupy substance that is formed in this way is first moulded into pellets and these are subsequently mixed with solvents to produce a clear, honey-like liquid. To form the solid transparent sheet which will become the basis of the film a constant flow of this liquid is spread in a uniform layer onto a turning wheel. As the solvents evaporate and the film dries, a fine continuous sheet is produced.

This base layer is then coated with a very fine emulsion of gelatin (derived from animal bones and skins) in which light-sensitive silver halide salts are suspended. Colour films need several successive layers of different emulsions and additional colour-forming chemicals to capture the image in view.

Most of us receive our photographs as coloured prints on paper and this medium, of course is derived from plants as well.

It is estimated that around 70 per cent of the adult population in Britain buy film regularly, spending about £400 million each year on this and in getting their films developed. Though chemicals control the developing processes, the images we capture with our cameras could not be reproduced as slides or snapshots without trees!

SPORTING PLANTS

The distinctive sound made by a leather-covered ball hitting a cricket bat is certainly a heavenly one to many of the game's most ardent devotees. Its distinctiveness is caused of course by the materials from which the bat and ball are made.

As almost every serious cricketer and fan will know, the most important piece of cricket equipment starts its life growing on a river bank or piece of marshy ground. Willow trees have given countless generations of cricketers the chief material for making bats and their wood is still in great demand today.

Of the nineteen willow species native to Britain (where the rules of cricket were laid down in 1744) only a variety of white willow, *Salix alba* var. *caerulea* has come to be considered as the best material for bats, and has thus assumed the name cricket bat willow. Before the introduc-

CRICKET BAT WILLOW

Salix alba var. *caerulea*

A very useful natural attribute of the cricket bat willow is its ability to grow exceptionally fast. Traditionally planted along low-lying fields and river banks in Essex, Suffolk and Norfolk, the trees are usually felled at 12 to 15 years old, by which time they may have grown to 20 metres or more in height.

For cricket bat production the trunk is first sliced into 'rounds' of about 28 inches (70 cm) in length and these are split into rough triangular 'clefts' in a process known as 'riving'. After careful seasoning, the bat's blade is finely cut to shape by hand and compressed for extra strength before being fitted to the cane and rubber handle.

One of several varieties of white willow, the cricket bat willow tree can be distinguished by its purple twigs and blue-grey, finely tapering leaves. Its bark is dark grey in colour and develops a network of thick ridges.

Though naturally of prime importance for their timber, the trees like many other willows are of great value to the countryside since their roots help bind the soil of river banks and reduce erosion.

tion of this hybrid to Suffolk in around 1780 and its subsequent adoption by cricketers, a variety of other woods were used. Early bats were rounded at the bottom very much like hockey sticks, tying in with the suggestion that the game itself originated among shepherds who used their crooks to hit the ball. The word cricket is indeed said to be derived from the Anglo Saxon 'cricce', which meant a shortened crooked stick.

Though the wood of the Kashmir willow is also used today as an alternative (this is the same variety of white willow as the cricket bat willow, but grown in southern India) — the toughness, resilience and comparative lightness of the English cricket bat willow are unique. Kashmir willow, which puts on two bursts of growth in one season because of the climate, tends to be harder and more brittle, and is therefore less likely to give such a good performance over the same period of time.

A bat's blade has to withstand tremendous punishment during the course of a professional match and absorb the impact of a ball travelling at anything up to 97 mph! It is cut from a triangular debarked wedge or 'cleft' of solid willow wood which has first been seasoned for 8 to 12 months.

After its initial rough shaping using a circular saw, the embryonic blade is left for another period to mature, during which the first stage of grading is carried out. The blade will then be carefully cut and individually shaped by hand, a critical process requiring a high degree of skill to ensure that the finished article is perfectly balanced.

To increase its strength and to avoid scarring when the ball hits the bat, the blade is compressed in a special machine which will also reveal any hidden faults or weaknesses that may be present. The 'v' shaped slot is then made in the neck of the blade ready for the handle to be inserted. After sanding and polishing and once the handle is in place, some blades may be finished with a fine cloth covering, whilst others are coated with a white or clear poly-

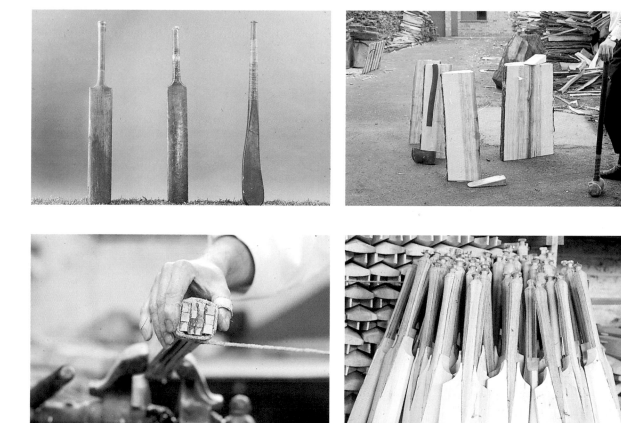

Early cricket bats were rounded at the base, rather like hockey sticks. Bats are manufactured from debarked wedges of seasoned cricket bat willow. Initially this is roughly shaped with a circular saw, and is left to mature before being cut and shaped by hand. A 'v' shape is cut in the neck of the blade for the handle which is made of several separate pieces of rattan cushioned with rubber. After sanding and polishing the blade can be covered with fine cloth, coated with a polyurethane film or treated with linseed oil.

urethane film. Traditional 'natural' blades however receive no other treatment than a coat of raw linseed oil, which must be applied to the face, edges, toe and back. This drying oil, pressed from flax seeds, sets to a hard elastic film and helps keep the blade in good condition.

The handle of the cricket bat is also made of specially adapted plant materials — but not of solid wood. The need for an exceptional capacity to absorb shock and diminish the impact of the ball led to the development of handles made from up to sixteen separate pieces of rattan cane (the sinewy stems of climbing tropical palms from South East Asia) cushioned with rubber. The canes are split, shaped and glued to make slabs, and these have rubber layers inserted between them. After drying and further shaping, the blade end of the handle is spliced into the body of the bat whilst the protruding part is bound with twine and fitted with a rubber grip. Together the shock absorbing layers of rubber and cane form a springy handle that will greatly help protect the batsman's hands.

ASH

Fraxinus excelsior

The toughness, resilience, straight grain and good bending properties of ash have made it exceptionally useful to manufacturers of sports equipment. Hockey stick heads, some squash racket frames and baseball bats are still made from ash of various thicknesses. The common or European ash is one of some 65 species all native to the Northern hemisphere, many of which yield a pale yellow wood of commercial importance. The tall, graceful tree reaches a height of 30 to 42 metres on good soil and its greenish-grey bark develops deep fissures with age.

In pre-Christian times the ash was worshipped in Scandinavia as a symbol of life, and was believed to have medicinal as well as mystical properties. According to Norse mythology, the most powerful god, Odin, carved the first human being from a piece of ash wood.

Willows, rubber trees and rattans are not the only plants essential to the game of cricket. English ash is used to make the stumps and bails and a central core of cork (cut from the cork oak) surrounded by layers of oiled wool and worsted yarn and further layers of cork, make up the inside of the leather-covered ball.

Despite the advances of modern technology, which have provided more resilient materials to match the new demands and greater physical strength of modern sportsmen and women, some traditional materials are still a vital part of certain sports.

Many of the woods that first became established, such as ash and hickory, are still in use because of special properties such as strength and lightness, the ability to resist repeated impact without splitting and to absorb shock. Though modern tennis racket frames are made of various combinations of carbon fibre, fibreglass and ceramic fibre or of aluminium alloys, in the early days of the game these were made of solid ash. By the 1930s lighter laminates had come to be used, reducing not just the weight of the racket but the tendency of the frame to distort. Until the last ten years or so, when the larger headed rackets began appearing on the courts, these laminate frames, which comprised 7 or 8 layers of ash and/or other woods such as beech and the tropical timber maroti, were still used. Decorative wedges of walnut, sycamore or mahogany sometimes filled the area between the handle and the frame. But the need for a much stronger, lighter frame able to take the higher string tension of the larger head necessitated the move away from wood in all but the cheapest non-professional rackets.

Conversely, whilst the cheaper baseball bats are now made of aluminium, those preferred by professionals — which can send the ball flying at the highest speeds — are made of solid ash. Though most competition squash racket frames are now made of man-made materials, a number of traditional frames are manufactured from a combination of four different woods selected from a range of six: ash, hickory, mahogany, beech, sycamore and obeche. Like much other modern sports equipment, including ash baseball bats, many of these wooden rackets are manufactured in the Far East.

Polo stick heads and skis were once made of solid ash. For polo sticks which still have handles made of rattan cane (as do hockey sticks), ash has now mostly been replaced by tipa wood (*Pterogyne nitens*) from Argentina. Top quality ski-makers have also largely abandoned ash, in favour not of other woods but of the modern very tough but lightweight

Lacrosse sticks are made from hickory, chiefly shagbark hickory. Professional baseball bats are made of ash, and this used to be the chief wood for polo stick heads, but has mostly been replaced by tipa wood. Hickory and ash are very strong woods which can be bent into shape whilst retaining their toughness

composite materials. Manufacturers of cross-country skis in Japan, however, do use laminates of hickory wood since it is able to withstand flexing and abrasion without splitting.

Hickory is, like ash, extremely strong and can be steam bent into various shapes whilst still retaining its toughness. Lacrosse sticks are still made from hickory, chiefly *Carya ovata* (shagbark hickory), native to the United States and imported into England from Alabama. To withstand the strain of play, a single length of wood is used and carefully bent into shape with the aid of steam. The wood is stiff and very dense — some 15 per cent heavier than ash and superior to it in its ability to withstand sudden impact.

SHAGBARK HICKORY

Carya ovata

One of the most important hickories for commercial use, the shagbark hickory is the source of very strong, impact-resistant timber. It is used amongst other things, for lacrosse stick manufacture and when burnt, for smoking foods. Like its famous relative the pecan (*Carya illinoiensis*) grown for its delicious nuts, the shagbark is native to North America, but has a wider natural distribution — from Quebec to Texas.

PERSIMMON

Diospyros virginiana

Though its own timber is very pale, the persimmon belongs to the same genus as trees whose heartwood yields the famous black ebony. There are about 200 different *Diospyros* species — mostly tropical and evergreen — and several, like persimmon, produce an edible fruit. *Diospyrus virginiana* is native to eastern and central North America. Its off-white sapwood timber has a fine, even texture and straight grain and is only used for specialist purposes. Since the wood is very dense — about 15 per cent heavier than beech — it is exceptionally hard, strong and impact-resistant and has long been favoured as a material for golf-club heads.

BILLIARD BALLS FROM COTTON

The first successful plastic ever made took the form of a billiard ball! In 1868, an American scientist, J.W. Hyatt, won a $10 000 prize for successfully making a plastic billiard ball from a mixture of camphor and nitro-cellulose — a substance made by reacting cotton linters, themselves almost pure cellulose, with acids.

'Now for a cannon, Bob!' Lithograph from 1820

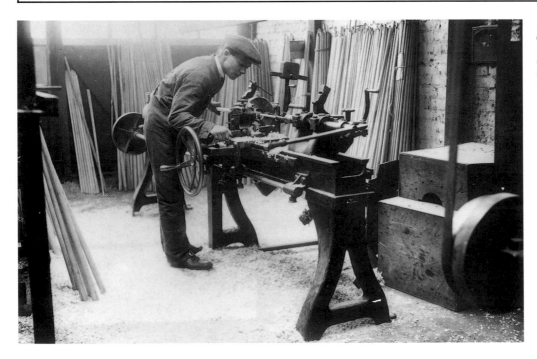

Billiard cue butts are made from rosewood or ebony, and the shafts from ash, maple or ramin

The shafts of most golf-clubs are now made of steel, but hickory was a favourite early candidate and some putters do feature hickory or ash. Carbon fibre has generally replaced wood for golf-club heads but manufacturers still use solid persimmon (*Diospyros virginiana*) or maple laminates for some. Persimmon is an exceptionally hard wood, well suited to the job it must perform. It is used essentially by traditionalists who prefer the feel of wood.

The green baize of the billiard table is also graced by wood, since the cues wielded by amateurs and professionals alike are made of nothing else. A cue comprises two main pieces, the butt or handle, and the shaft. The butt is normally made from rosewood (*Dalbergia* spp.) now grown in Indonesia or ebony (*Diospyros* spp.) from India and Sri Lanka, whilst the shaft will generally be made of ash (for the best quality cues), maple (for the next best) and ramin, from South East Asia (for the cheapest). Ramin is currently being imported into Britain in large quantities for a large number of joinery applications and for a range of miscellaneous goods from broom handles to set-squares.

In between the butt and the shaft of the billiard-cue a coloured piece of veneer, often of sycamore or maple, about 0.25 cm thick is inserted to create an attractive figuring as the cue is turned.

The roll of plants!

A tropical wood well known for its association with one of the oldest British games is lignum vitae. Flat green and crown green bowls were once made almost exclusively from this dense wood, the heaviest timber in commercial use. A shortage of lignum vitae of the right size due to overcutting has led to the widespread use of synthetic materials, but none the less, some timber still ends up serving this long-established purpose.

LIGNUM VITAE

Guaiacum officinale

The best commercial lignum vitae is said to come from the West Indies and coastal regions of Colombia and Venezuela, but trees belonging to the genus *Guaiacum* from which the wood is taken, are also native to Central America.

The small, tropical trees were given the name lignum vitae ('wood of life'), because at the time of their discovery in the sixteenth century, their resin was thought to be a cure for several diseases. Their remarkable timber — a distinctive greenish black in colour, and very rich in resin — became and has remained a very valuable commodity for its tremendous hardness, strength and density. The fine-textured, closely interlocked grain, which makes lignum vitae 70 to 80 per cent heavier than oak, combines with its resin content to make the wood waterproof and exceptionally resistant to rot and abrasion.

BALATA
Manilkara bidentata
The latex tapped from wild balata trees is valued for its special non-elastic properties when reacted with other compounds and set hard.

Balata trees which are native to most of Tropical America grow to a height of around 40 metres and are usually tapped three times a year. As well as being used to coat golf balls, the latex is sometimes substituted for chicle gum (*Manilkara zapota*), the chief ingredient of chewing gum.
The best golf balls are coated with balata latex

Bowls are made in sets of four (or two for crown green bowls) and they must all be cut from the same log so that their specific gravity is identical. It is essential that their weight and bias should be exactly equal so that their movement on the green does not vary. No two sets of bowls are quite alike and the player must get to know their different characteristics. Lignum vitae logs are chosen for straightness, with their inner pith always forming the horizontal axis of the bowl. Only seasoned heartwood is used and any cracks that may appear are carefully filled before the bowl is turned and accurately shaped. When not in play bowls must be kept in a cool place to prevent cracks from developing.

Some rather faster moving balls essential to another sport that is played on grass also incorporate materials made from tropical plants. Top quality golf balls are currently made partly from balata rubber, a natural latex exuded from a number of trees belonging to the genus *Manilkara*, but chiefly *M. bidentata*, native to Trinidad and parts of South America. The inner core of a golf ball is made from an enormous length of tightly bound elastic about 1800 metres long. Though the covering which surrounds this mass of elastic in 95 per cent of all golf balls made is a synthetic thermoplastic, most of the world's professional golfers prefer to play with those covered with balata. Only a very thin layer is applied, but this is enough to give the ball its distinctive qualities and feel. The main advantage of balata is that it has much more back-spin than its alternatives, meaning essentially that once the ball hits the ground and begins to roll forward, it counteracts this movement, giving the player much more control. Whilst in the air the balata-covered ball spins a lot faster than those coated with thermoplastic, but it is likely to be damaged far more easily by amateur players who do not hit it quite so cleanly.

PLANTS INSTRUMENTAL TO GOOD MUSIC

We will never know whether the earliest musical instrument on earth was made when someone hit or blew air into a naturally resonant material and was fascinated by its sound. And was it bone or wood or maybe animal skin or sinews stretched taut that gave the first 'musical' note? Certainly all sorts of animal products have been used for making music round the world, from jaguar bone flutes and conch shell trumpets to armadillo-backed guitars. But plants have provided us with a vast pool of natural resources for music-making.

It would be impossible to list all the many kinds of wood that have been hollowed out or fashioned into drums (and drumsticks) world-wide, and almost as difficult to name the plants that have been used for making simple woodwind instruments. Plants with naturally hollow stems such as bamboo and reeds, however, have been the natural choice for many peoples where these materials have been available, and some of our loveliest and simplest contemporary music is produced by them.

A whole range of flutes and panpipes are made and used by Andean peoples of Peru, Bolivia and Ecuador from native bamboo species such as those belonging to the genus *Chusquea*, and from reeds such as *Arundo donax*. The light airy sounds made by the tiniest sikus or zampoñas, and extraordinary deep and husky bass notes of the toyos from Bolivia, are produced from different lengths and widths of hollow stem.

The reed *Arundo donax* has been of particular importance to Western music too. European panpipes made in France and the Balkan countries have traditionally used its stems, gathered from the Danube delta area. This same plant has given us, as the name suggests, the best 'reeds' for clarinet and oboe mouthpieces, currently collected from various districts in Southern France.

For wind instruments of the East, bamboos have been especially significant since the earliest times. The national instrument of Japan, the shaku hachi, is made from the lower end of small-culmed bamboos, while the Chinese xiao, a notched flute played in modern orchestras is made from the bamboo *Phyllostachys nigra*. Each of the 950 pipes belonging to an unusual organ built in the Philippine city Las Piñas in 1818 is a hollowed bamboo culm and the instrument is said to be still in working order.

Bamboo has many other musical uses too. In West Java more than twenty percussion, wind and string instruments are made from various species but mostly from the genus *Gigantochloa*. In Bali the fungklih xylophone has bamboo keys, whilst others have resonators made of this material.

When it comes to the different tree species used to make the world's musical instruments the picture becomes much more complex. Locally grown or native trees have naturally provided the basic raw materials for most of the instruments played by different peoples, but where more suitable foreign timbers have become available these have often been adopted.

Chusquea, *used for making 'bamboo' flutes in South America*

The shaku hachi, the national instrument of Japan, is made from bamboo

EBONY
Diospyros spp

Distinctive black ebony wood, valued for centuries for cabinet work and decorative inlays, is used today to make components of a variety of musical instruments and some sports equipment. Organ stops and black piano keys, fittings for violins, guitars and bagpipes, as well as billiard cues and castanets all utilize this hard, heavy wood. The most important ebony species for commercial use are *Diospyros ebenum* from India and Sri Lanka and *D. reticulata* from Mauritius, though supplies are now also exported from West Africa.

New wooden bagpipes being tested!

Bagpipes, for example, which were introduced to Scotland in the thirteenth century from Central and Southern Europe, would have made use of temperate trees. Today they may be made of ebony (*Diospyros* spp.) from various tropical countries, African blackwood (*Dalbergia melanoxylon*) from Tanzania or shee-sham wood (*Dalbergia* spp.) from Pakistan.

Shee-sham, a kind of rosewood, is the material used to make the neck and pegs of the jowzé, a traditional Iraqi bow-stringed instrument which has a resonator made from half a coconut, and sound box made from the amniotic sack of a goat! This 'spike fiddle' as it is also known is currently played in classical ensembles in Iraq.

Many different woods are used for various orchestral instruments

Rosewoods of various kinds are much in evidence amongst many instruments made with wood. Acoustic and electric guitars both incorporate them today, as do many violins. African blackwood is also very popular. Most modern clarinets are made almost entirely of this wood as are oboes. Recorders meanwhile are utilizing box wood and satinwood — both valued for their fine golden-yellow finish — whilst bassoons often feature sycamore or maple.

As is true for the manufacture of any specialist equipment, the production of musical instruments to a consistent standard depends not just on how they are put together, but also on the particular properties of the materials used.

A piano, for example, is made up of a large range of different woods, each one selected for the special function it will perform. Whilst the white keys will often be made of basswood (*Tilia americana*) and the black of ebony — both hard, stable timbers that resist distortion — the soundboard and its ribs require a lightweight wood with a high degree of elasticity, enabling it to vibrate and give resonance to the strings. Spruce has been an ideal timber for this purpose, though other light woods are sometimes substituted. The hammer heads that strike the strings may be made of hornbeam (*Carpinus* spp.), sapele (*Entandrophragma cylindricum*) or mahogany, and the main beam of North American rock maple (*Acer saccharum*).

Violins, violas, cellos, and other instruments of the viol family likewise comprise various kinds of wood, chosen for their different characteristics. Since their top side or belly has similar requirements to a piano soundboard, it is invariably made of a light coniferous wood — usually spruce. The backs, sides, necks and scrolls meanwhile are traditionally made of sycamore or maple. The distinctive rippled grain used for the backs has become so much associated with violin-making that it is commonly referred to as 'fiddle back'.

FIELD MAPLE

Acer campestre

The rippled grain traditionally used to make the backs of violins is a distinctive feature of maple wood. The common field maple, native to Britain, was once an important source of this decorative wood but most trees of useful timber-producing size have now been felled. There are, however, some 200 different maple species in existence, widespread in north temperate regions, and a number are now used by woodworkers for a variety of string and woodwind instruments.

Maple wood, used to make the back, sides and neck of violins has a distinctive rippled grain (left), commonly referred to as fiddle-back

The master violin maker Antonio Stradivarius (1644–1737) was the first to use a bridge of maple to support the strings of violins. The quality of his instruments, however, was determined to a large degree by the way in which the wood was seasoned and the composition of the finishing varnish.

Some of the best old Italian cellos are backed with either poplar or willow wood though beech or one of the fruit woods is sometimes utilized. Nowadays ebony is used for the finger board and tailpiece, and two smaller parts — the topnut and the saddle — of many violins. Tuning pegs are often cut from Honduras rosewood (*Dalbergia stevensonii*), a very high density timber also commonly selected for high quality xylophones, though boxwood is considered smarter.

That now ubiquitous Moorish invention, the guitar — an essential part of the identity of Spanish gypsies, and in electronic form, young people everywhere — still relies on wood to a great extent. Spanish or acoustic guitars have used a range of different woods and wood veneers over the centuries. Today sycamore, rosewood and maple are very popular with manufacturers. Ninety-nine percent of all electric guitars are made essentially of wood. The most modern might have a body comprising mahogany for the back piece, a heavy wood which gives a warm, deep resonance to the sound — and maple for the front, a lighter wood which gives a 'brightness' to the strings' vibration. Alder, ash and maple may be used too according to the personal preference of the player. Fingerboards are often cut from pieces of ebony or rosewood, or bubinga (*Copaifera salikounda*) a heavy timber from West Africa that is similar to rosewood in appearance, with a distinctive highly decorative colouring and grain.

Whatever the instrument we play or listen to, musicans and music lovers everywhere will surely agree that plants have played a vital part in helping us create the sounds which both identify our different cultures and unite us all.

Guitars are made from many different woods. This one features mahogany, with an ebony fingerboard

Index

Abies spp. firs, 107, 189
Abies alba silver fir, 110
Acacia spp. wattle, 9, 55
Acacia catechu Indian tree, 24, 61
Acacia coriacea wattle, 97
Acacia mearnsii black wattle, 56, 57, 60, 180
Acacia senegal, see gum arabic
Acer campestre field maple, 221–3
Acer pseudoplatanus sycamore, 116, 214, 217, 221, 223
Acer saccharum rock/sugar maple, 77, 78, 175–6, 221
Adansonia digitata baobab, 38
Adansonia gregorii 97
Aegilops spp., 70
Aeschynomene aspera, 65
Aeschynomene indica, 65
Afzelia bipindensis, 176
Afzelia pachyloba, 176
agar, 45, 102, 146
Agathis spp., 204
Agave spp. sisal, 62, 186–7, 188, 189
Agave fourcroydes henequen, 188
Agave pacifica maguey, 44
Agave sisalana sisal, 188
Aids, 166–7
aircraft, 170, 180–2
Alaria spp., 20
Aleurites spp. tung oils, 126, 203
algae, 103, 191
 see also seaweeds
alginates, 127, 139, 170, 180
 sizes, 45, 125, 202
 thickeners, 14, 18, 20, 45, 58
Alkanna lehmannii alkanet, 120
allergies, 148
Allium spp., 144
Allium cepa onion 58, 144
almond, bitter *Prunus dulcis* var. *amara*, 17, 28
almond, sweet *Prunus dulcis* var. *dulcis*, 17, 67
Aloe vera, 20, 21
aloes, 153
Amaranthus spp., 84
Amyris balsamifera West Indian sandalwood, 33
anaesthetics, 153–7, 159
Ananas comosus pineapple, 53, 99, 153
Aniba rosaeodora, 115
Anisoptera spp. mersawa, 177
annatto *Bixa orellana*, 23, 75

Annona muricata soursop, 99, 102
Anogeissus latifolia gum ghatti, 190
antibiotics, 144
antiseptics, 139–40, 141
apple *Malus sylvestris*, 116, 119
apple-berry *Billardiera scandens*, 97
apricots, 8, 17, 67
arrowroot *Maranta arundinacea*, 37, 44
Artabotrys odoratissimus ylang-ylang, 30, 31
arthritis, 146–7
Artocarpus spp. breadfruit trees, 183
Artocarpus altilis breadfruit tree, 36
Arundo donax reed, 130, 219
ash *Fraxinus excelsior*, 214
 construction and joinery, 116, 176, 179, 181, 223
 sports equipment, 215–17
aspen *Populus tremula*, 54, 127, 199
Astrocaryum chambira Chambira palm, 35
Atropa belladonna deadly nightshade, 149
Attalea funifera piassava, 128
atuqsara *Phytolacca bogotensis*, 6
avocados, 16, 17, 99

Bactris spp., 35
Balanites aegyptiaca, 6
balata trees *Manilkara bidentata*, 218
balsa *Ochroma lagopus*, 108, 181, 183, 186
balsam, 29, 192, 193
bamboo *Bambusa* spp., 62
 building and construction, 129, 133, 181, 183
 musical instruments, 218, 219, 220
 pulp, 56, 196, 201
bamboo, giant *Dendrocalamus* spp., 183
bananas, 67, 99, 192
bandages, 139
baobab *Adansonia digitata*, 38
bark cloth, 36–9
bark paper, 198
barley, 81, 87, 101
basswood *Tilia americana*, 221
bath oils, 10–11, 13
bean, winged *Psophocarpus tetragonolobus*, 102
beans, 67, 99

beech, European *Fagus sylvatica*, 2, 117
 construction and joinery, 116, 117, 214, 223
 pulp, 9, 54, 127, 199
beers, 5, 82, 83, 203
Beta vulgaris sugar beet, 10, 77, 79, 192
Betula spp., *see* birch
Bignonia chica, 23
bilberry *Vaccinium myrtillus*, 96
Billardiera scandens apple-berry, 97
billygoat plum *Terminalia ferdinandiana*, 97
birch *Betula* spp., 116, 117
 construction and joinery, 107, 111, 179, 181, 182
 pulp, 54, 127, 199
birch, common *Betula pubescens*, 117
birch, silver *Betula pendula*, 117
Bixa orellana annatto, 23, 75
blackwood, African *Dalbergia melanoxylon*, 220, 221
blood pressure, 162–3
bloodroot *Sanguinaria canadensis*, 160
boats and ships, 182–9
Boehmeria nivea ramie, 48, 49
 var. *tenacissima*, 49
Bombyx mori silkworm, 50–2
Boswellia carterii frankincense, 27, 28
box *Buxus sempervirens*, 116, 176, 210, 221, 223
boxwood, maracaibo *Casearia praecox*, 210
brazilwood *Caesalpinia* spp. 58
bread, 68–70
bread, laver *Porphyra umbilicalis*, 102
breadfruit trees *Artocarpus* spp., 36, 183
brooms, household, 128
Brosimum galactodendron cow tree, 71
Brosimum utile cow tree, 71
Broussonetia papyrifera paper mulberry, 36, 37, 38, 196
Brugmansia spp. datura, 137
buckthorn *Rhamnus purshiana*, 153
buckwheat *Fagopyrum sagittatum*, 84
buildings, 105–19, 129–33
bulrush *Schoenoplectus lacustris*, 123
Butea frondosa, 120, 208
butter, 71–3

Buxus sempervirens box, 116, 176, 210, 221, 223

Caesalpinia spp. brazilwood, 58
calabar bean *Physostigma venenosum*, 150
Calamus caesius rattan, 122
Camellia sinensis tea plant, 89
 var. *assamica* Assam tea, 92
 var. *sinensis* China tea, 92
camomile, 1, 8
 German *Chamaemelum recutita*, 26
 Roman *Chamaemelum nobilium*, 26
camphor *Cinnamomum camphora*, 142
Cananga odorata ylang-ylang, 30, 31
Canarium schweinfurthii, 131
cancer, 145, 158–61
Cannabis sativa, see hemp
canvas, 49, 62
Caralluma speciosa, 30
cardamom, 29, 98
Carludovica palmata toquilla, 65
carob, *see Ceratonia siliqua*
carpets, 125
Carpinus spp. hornbeam, 221
Carpobrotus spp. pigface fruits, 97
carrageenan, 18, 58, 61, 102, 146
carrots, 16–17
Carya spp. hickory, 181, 214, 215, 217
Carya illinoiensis pecan, 215
Carya ovata shagbark hickory, 215
Caryocar brasiliense pequi fruits, 23
Casearia praecox maracaibo boxwood, 210
cassava/manioc *Manihot esculenta*, 192
 food, 67, 99, 100, 101
 starch, 44, 45, 58
Cassia angustifolia Indian senna, 152
Cassia senna Alexandrian senna, 152, 153
Castanea dentata, 61
Castanea sativa sweet chestnut, 61, 116, 179–80
Castanospermum australe Moreton Bay chestnut, 166, 167
Castilla elastica, 173
castor oil, 1, 191
 cosmetics, 17, 24, 28
 medical usage, 153, 157

castor oil plant *Ricinus communis*, 10
Catharanthus roseus rosy periwinkle, 159–60
cedar, incense *Libocedrus decurrens*, 208–9
cedar, pencil *Juniperus virginiana*, 208
cedar, Western red *Thuja plicata*, 112, 113
cedar wood, 28, 30, 185, 186
Ceiba pentandra kapok tree, 52–4
Cephaelis ipecacuanha ipecac, 142, 143
Ceratonia oreothauma, 97
Ceratonia siliqua carob/locust bean, 45, 97
 gum, 18, 58
cereals, 81–4
Chamaemelum nobile Roman camomile, 26
Chamaemelum recutita German camomile, 26
Chamaerops humilis, 120
Chenopodium quinoa quinoa, 84
cherrywood, 58, 176
chestnut, Moreton Bay *Castanospermum australe*, 167
chestnut, sweet *Castanea sativa*, 61, 116, 179–80
chewing gum, 98, 218
chewing sticks, 15
chicle tree *Manilkara zapota*, 98, 218
chicory *Cichorium intybus*, 87
chipboard, 111
Chlorophora excelsa iroko, 113, 183, 186, 189
chocolate, 68, 98
 substitute, 97
Chondrodendron spp., 155–6, 159
Chondrodendron tomentosum, 157
Chondrus crispus Irish moss, 18, 62
Chusquea spp., 219
Cichorium intybus chicory, 87
Cinchona spp., 159, 163–6
Cinchona calisaya, 165
Cinchona ledgeriana, 164, 165
Cinchona officinalis, 165
Cinchona succirubra, 165
Cinnamomum camphora camphor, 142
cinnamon, 28, 30, 98, 141
Citrullus lanatus tsamma/karkoer, 96
Citrus aurantium Seville orange, 76, 77
Citrus bergamia Bergamot orange, 76
Citrus grandis pummelo, 76
Citrus limon lemon, 7, 30, 76, 141
Citrus sinensis sweet orange, 76–7

Citrus × paradisi grapefruit, 76
cloudberries *Rubus chamaemorus*, 96
clovers *Trifolium* spp., 73
cloves, 15, 98
coca bush *Erythroxylum coca*, 153–5
cocaine, 153–4
cocksfoot *Dactylis glomerata*, 73
cocoa *Theobroma cacao*, 18, 98
coconut palm, *see* palm, coconut
codeine, 145
Coffea arabica, 85, 86
Coffea canephora, 85
coffee, 84–8
coir, 3, 125, 187
Coix lacryma-jobi Job's tears, 36, 84
Colchicum autumnale autumn crocus/meadow saffron, 147
colds and coughs, 141–3
Colocasia esculenta taro, 37–8, 99
 var. *antiquorum*, 44
colours, cosmetic, 21–6
 see also dyes; pigments
Commifera abyssinica, 27
Commifera myrrha, 27, 29
contraception, 149–52
Convolvulus brasiliensis, 38
Copaifera spp., 192, 193, 204
Copaifera salikounda bubinga, 233
copal, 204
Copernicia prunifera Brazilian wax palm, 24, 25, 120, 180
copra, 2, 3
Corchorus spp. jute, 124, 125, 187
Cordeauxia edulis yeheb, 97
Cordia sebestena, 38
coriander, 28, 30
cork, 126, 214
corkwood tree *Duboisia* spp., 149
corn, *see* maize
cosmetics, 16–26
cotton *Gossypium* spp., 35, 183, 216
 clothing, 41–4, 54
 medical usage, 138, 139, 140
 paper making, 196, 201
 thickeners, 8–9, 13
cotton, sea island *Gossypium barbadense*, 41, 43, 44
cotton, upland *Gossypium hirsutum*, 42, 43–4
cottonseed oil, 2, 73, 75, 150
cow trees *Brosimum* spp., 71
crayons, 208–9
cricket, 211–14
crocus, autumn *Colchicum autumnale*, 147
Croton tiglium, 160
crowberries *Empetrum nigrum*, 96
curare, 155–7

Curcuma longa turmeric, 38
Cyamopsis tetragonolobus guar gum, 18, 19, 58, 190, 202
Cymbopogon schoenanthus desert camel grass, 27
Cyperus pangorei, 124
Cyperus papyrus papyrus, 184, 187, 195, 196
Cyperus tegetiformis Chinese sea grass, 124
cypress oils, 32

Dactylis glomerata cocksfoot, 73
Daemonorops spp. rattan, 122
Dalbergia spp. rosewood, 216, 217
 joinery timber, 111, 114, 115, 116
 musical instruments, 220, 221, 223
Dalbergia cearensis tulip tree, 181
Dalbergia melanoxylon African blackwood, 220–1
Dalbergia stevensonii Hondurus rosewood, 223
dammar *Shorea* spp., 204
dandelion *Taraxacum bicorne*, 173
dandelion *Taraxacum officinale*, 58, 87
datura *Brugmansia* spp., 137
deadly nightshade *Atropa belladonna*, 149
Dendrocalmus giganteus giant bamboo, 183
Dendrocalmus strictus giant bamboo, 183
detergents, 8, 14, 57
Digitalis spp. foxgloves, 157–8
Digitalis lanata, 158
Digitalis purpurea, 158
Dioscorea spp. yams, 99, 149–50
Dioscorea deltoidea, 149
Dioscorea nipponica, 149
Dioscorea tokoro, 149
Dioscoreophyllum cumminsii serendipity berry, 80
Diospyros spp. ebony, 216, 217, 220, 223
Diospyros canaliculata, 131
Diospyros ebenum, 220
Diospyros reticulata, 220
Diospyros usambarensis, 15
Diospyros virginiana persimmon, 216, 217
Diospyros whyteana, 15
Dipterocarpus spp. keruing, 177
dock construction, 189
drawing instruments, 210
Duboisia spp. corkwood tree, 149
Dyera costulata, 98
Dyera lowii, 98
dyes, 38–9, 57–9, 75, 205–7
 hair, 24–6
 see also colours, cosmetic; pigments
dynamite, 191

Eau de Cologne, 30, 32
ebony, *see Diospyros* spp.
ekki *Lophira alata*, 176, 189
Elaeis guineensis see palm, African oil
Eleusine coracana finger millet, 82
Empetrum nigrum crowberries, 96
Enantia chlorantha, 131
engines, 190–3
Entandrophragma cylindricum sapele, 176, 221
Ephedra spp., 148
Epilobium angustifolium fireweed, 96
erasers, 209
Erythroxylum coca coca bush, 153–5
eucalyptus, 9, 55–6
 building and construction, 113, 177
 medical usage, 138, 141
Eucalyptus citriodora, 141
Eucalyptus fastigiata, 56
Eucalpytus globulus, 56, 141, 199
Eucalyptus grandis, 56, 199
Eucalyptus marginata jarrah, 113
Eucalyptus paniculata, 56
Eucalyptus saligna, 111
Eucalyptus sturtiana, 141
Euchlaena mexicana teosinte, 83
evening primrose *Oenothera* spp. 147

Fagopyrum sagittatum buckwheat, 84
Fagus sylvatica see beech, European
fennel, 15, 21
Ferula spp., 9
fescue, meadow *Festuca pratensis*, 73
feverfew *Tanacetum parthenium*, 21, 147
fibreboards, 111
fibres, synthetic, 54–5
Ficus spp. figs, 35, 36, 38
Ficus anthelmintica fig, 160
Ficus benghalensis Indian tree, 26
Ficus carica fig, 87
Ficus elastica India rubber tree, 173
Ficus tecolutensis, 198
Ficus tinctoria, 38
fir, Douglas *Pseudotsuga menziesii*, 107, 109, 110, 111, 112, 186, 189
fir, silver *Abies alba*, 110
firs *Abies* spp., 107, 189
fireweed *Epilobium angustifolium*, 96
flavour enhancers, 80
flavourings, 14–15, 17, 165
flax *Linus usitatissimum*, 46–7, 111, 187

paper making, 196, 201
see also linen; linoleum; linseed oil
flax, New Zealand *Phormium tenax*, 39–40, 197
floor coverings, 40, 123–6
foxgloves *Digitalis* spp., 157–8
frankincense *Boswellia carterii*, 27, 28
Fraxinus excelsior, see ash
Fucus spp., 20, 22
fuels, engine, 192–3
Funtumia elastica, 173
furniture, 40, 114, 115–23

gamboge *Garcinia hanburyi*, 205–6, 207
Garcinia mangostana mangosteen, 102
Garcinia morella, 207
gardens, fragrant, 29
garlic *Allium sativum*, 144
Gelidium spp., 45
Genipa americana, 23
Gigantochloa spp., 219
Gigartina stellata Irish moss, 18, 62
Gloiopeltis furcata, 45
glycerine, 5, 13, 18
Glycine max, see soyabean
Glycyrrhiza glabra liquorice, 142
golf, 216–18
Gonystylus bancanus ramin, 210, 216, 217
Gossweilerodendron balsamiferum agba, 185
Gossypium arboreum, 43
Gossypium barbadense sea island cotton, 41, 43, 44
Gossypium herbaceum, 43
Gossypium hirsutum upland cotton, 42, 43–4
Gossypium raimondii, 43
Gossypium spp., *see also* cotton
grapefruit *Citrus* × *paradisi*, 76
grass, Chinese sea *Cyperus tegetiformis*, 124
grass, desert camel *Cymbopogon schoenanthus*, 27
grass, esparto *Stipa tenacissima*, 62, 187, 200
grass, lemon *Cymbopogon citratus*, 30
grass, Napier, 192
grass, satintail *Imperata brasiliensis*, 132–3
grass, timothy *Phleum pratense*, 73
grasses, 67, 68, 71–3
see also cereals
greenheart tree *Ocotea rodiaei*, 151, 185, 189
Grevillea, 97
groundnut, *see* peanut
Guaiacum officinale lignum vitae, 217–18

guar gum *Cyamopsis tetragonolobus*, 18, 19, 58, 190, 202
guayule *Parthenium argentatum*, 173
gum arabic *Acacia senegal*, 45, 146
cosmetics, 18, 22
dyeing and printing, 58, 203
paints, 127, 205
gum ghatti *Anogeissus latifolia*, 190
gum karaya, 15, 58
gum, locust bean, 18, 58
gum tragacanth, 18, 58
gums, 28, 58, 190
cosmetics, 18, 19
sizing agents, 45, 202
see also resins; rosin

Haematoxylum campechianum logwood, 58
hair conditioners, 9–10
hair dyes, 24–6
hairsprays, 26
Hamamelis virginiana wych hazel, 140
hardboard, 111
hats, 62–5
Helianthus annuus, see sunflower
hemlock trees *Tsuga* spp., 61
hemlock, western *Tsuga heterophylla*, 9, 56, 109, 110, 112
hemp *Cannabis sativa*, 48–9
paper making, 196, 201
ropes, 47, 48–9, 186, 187, 189
hemp, Manila *Musa textilis*, 186–9
henbane *Hyoscyamus* spp., 148, 149
henna *Lawsonia inermis*, 24–5, 28
heroin, 146
Hevea brasiliensis, see rubber tree
hickory *Carya* spp., 181, 214, 215, 217
hickory, shagbark *Carya ovata*, 215
hornbeam *Carpinus* spp., 221
humectants, 13–14, 18
hyacinth, 29
hyacinth, water, 192
Hymenaea spp., 204
Hyoscyamus albus, 148
Hyoscyamus niger henbane, 148, 149

ice-cream, 102
ichu *Stipa ichu*, 184
Imperata brasiliensis satintail grass, 132, 133
Indian tree *Acacia catechu*, 24, 61

Indian tree *Ficus benghalensis*, 26
indigo, 59
Indigofera tinctoria, 24
influenza, 144
inks, 45, 58, 203
see also dyes; paints
ipecac *Cephaelis ipecacuanha*, 142, 143
Ipomoea batatas sweet potato, 99, 192
iris, 59
Irish moss, 18, 62
iroko *Chlorophora excelsa*, 113, 183, 186, 189
Isatis tinctoria woad, 58

Jacaranda spp., 161
Jacaranda caucana, 160
jarrah *Eucalyptus marginata*, 113
jasmine, 8, 29, 30, 32
Jasminum sambac, 29
Jatropha spp., 160
jeans, 48
Job's tears *Coix lacryma-jobi*, 36, 84
jojoba *Simmondsia chinensis*, 7, 8, 24
Juglans nigra black walnut, 26, 178–9
Juglans regia English walnut, 178, 179
Juncus acutus, 124
Juncus maritimus, 124
Juniperus virginiana pencil cedar, 208
jute *Corchorus* spp., 124, 125, 187

kapok tree *Ceiba pentandra*, 52–4
kelp, giant *Macrocystis pyrifera*, 45, 125, 192
kelps *Macrocystis* spp., 19, 20, 139
Khaya spp., 115
Koompassia malaccensis kempas, 176

lac tree *Schleichera oleosa*, 10
lace bark tree *Lagetta lagetto*, 40
lacquer/varnish tree *Rhus verniciflua*, 120
Lagenaria breviflora, 151
Lagetta lagetto lace bark tree, 40
Laminaria spp., 45
Laminaria digitata, 19
Laminaria hypoborea, 19
Landolphia spp., 173
larches *Larix* spp., 112, 186
Lavandula angustifolia, 12
Lavandula vera, 12
lavender, 1, 11–12, 13, 30, 33
Lavigeria macrocarpa, 166
Lawsonia inermis henna, 24–5, 28

laxatives, 152–3
Ledum palustre labrador tea, 96
lemon *Citrus limon*, 7, 30, 76, 141
Libocedrus decurrens incense cedar, 208–9
lichens, 58, 96
lignum vitae *Guaiacum officinale*, 217–18
lime *Citrus aurantifolia*, 33, 76
lime *Tilia europaea*, 119
linen, 46–7, 181
medical usage, 139, 157
paper making, 196, 201
lingerie, 54–5
linoleum, 124
linseed oil, 46, 57, 120, 124, 213
cosmetics, 17, 24
paints, 126, 204, 205
Linum usitatissimum flax, 46–7, 111, 187
paper making, 196, 201
see also linen; linoleum; linseed oil
Lippia dulcis, 80
lipstick, 22, 24
liquorice *Glycyrrhiza* spp., 142, 143
Lithospermum erythrorhizon, 24
locust bean, *see* Ceratonia siliqua
logwood *Haematoxylum campechianum*, 58
Lolium multiflorum Italian ryegrass, 72, 73
Lolium perenne perennial ryegrass, 72, 73
Lonchocarpus cyanescens, 59
loofahs, 12–13
Lophira alata ekki, 176, 189
lubricants, 191
Luffa acutangula, 12
Luffa aegyptiaca, 12
Lupinus mutabilis tarwi, 102

Macadamia spp., 16–17
nut oils, 16–17
Macrocystis spp. kelps, 19, 20, 139
Macrocystis pyrifera giant kelp, 45, 125, 192
madder *Rubia tinctorum*, 58, 205–7
madder, wild *Rubia peregrina*, 205
magic, 23, 137–8
maguey *Agave pacifica*, 44
mahoganies, 185, 214
joinery timber, 114, 179, 182, 186
musical instruments, 221, 223
mahogany, Honduran, 185
mahogany *Swietenia* spp., 115, 116
maize/corn *Zea mays*, 13, 73, 82–3, 87, 192, 205

food, 67, 80, 81, 82–3, 99, 101
starch, 13, 22, 44, 58
malaria, 163–6
Malus sylvestris apple, 116, 119
mandrake, American
    Podophyllum peltatum, 160
mango, 99, 183
mangosteen *Garcinia
    mangostana,* 102
mangrove, 38, 56, 61
Manihot esculenta, see cassava/
    manioc
Manihot glaziovii Ceara rubber
    tree, 173
Manilkara bidentata balata trees,
    218
Manilkara zapota chicle tree, 98,
    218
manketti tree *Ricinodendron
    rautanenii,* 96
maple, field *Acer campestre,*
    221–3
maple, rock/sugar *Acer
    saccharum,* 77, 78, 175–6, 221
maples, 56, 216, 217
Maranta arundinacea arrowroot,
    37, 44
margarine, 71, 73–5
Mauritia flexuosa buriti palm,
    34, 35, 132
meadows, hay, 72
medicines, 134–53
    rainforests, 143, 150, 151,
    158–59, 161, 167
Megaphrynium macrostachyum,
    131
melon *Citrullus lanatus,* 96
Mentha spp. mints, 1, 14–15,
    29, 143
Mentha arvensis, 14, 15
Mentha cardiaca, 15
Mentha piperata, 14, 15
Mentha spicata, 14, 15
Metroxylon sagu sago palm, 44,
    67
millets, 81, 82
mints, *see Mentha* spp.
miraculin, 80
mistletoe, 118
Mora excelsa, 176
morphine, 145–6
Morus celtidifolia, 198
mulberry, black *Morus nigra,*
    50, 51
mulberry, paper *Broussonetia
    papyrifera,* 36, 37, 38, 196
mulberry, white *Morus alba,* 50
Musa textilis Manila hemp,
    186–9
musical instruments, 218–23
muslin, 43, 181
myrabolans tree *Terminalia
    chebula,* 61
myrrh *Commifera* spp., 27, 28

nappies, disposable, 56

narcissi, 22, 29
Nardostachys jatamansi
    Himalayan valerian, 27
Nauclea diderichii opepe, 186,
    189
neroli oil, 30
nettle, *Urtica dioica,* 8, 21, 48,
    58

oak, cork *Quercus suber,* 126,
    214
oak, pedunculate/common
    Quercus robur, 118, 119
oak, sessile/durmast *Quercus
    petraea,* 118
oak, valonea *Quercus aegilops,*
    61
oak timber, 118–19, 176
    construction and joinery, 105,
    106–7, 116
    ships and docks, 184, 186, 189
    oaks, 26, 118
    acorns, 87, 126
    pulp, 9, 56
    silkworm larvae food, 51, 157
oar weeds *Laminaria* spp., 19,
    20
obeche *Triplochiton scleroxylon,*
    131, 185, 214
oca *Oxalis tuberosa,* 100, 102
Ochroma lagopus balsa, 108, 181,
    183, 186
Ocotea rodiaei greenheart tree,
    151, 185, 189
Ocotea venenosa, 155
Oenothera spp. evening
    primrose, 147
oil palm, *see* palm, oil
oils,
    cosmetics, 16–18, 24
    engine lubricants, 190–2, 193
    food, 71, 73–5
    inks, 203, 204
    medicine, 141–2, 146, 147, 150
    paints, 126, 204
    perfumes, 28–33
    toiletries, 1–13
    see also castor oil; linseed oil;
    sunflower seed oil
olive oil, 2, 16–17, 28, 75
onion *Allium cepa,* 58, 144
opepe (*Nauclea diderichii*), 186,
    189
orange, perfumery, 30, 32
orange, Bergamot *Citrus
    bergamia,* 76
orange, Seville *Citrus
    aurantium,* 76, 77
orange, sweet *Citrus sinensis,*
    76–7
orchid, white dove, 32
Oryza spp., *see* rice
Osyris tenuifolia, 33
Oxalis tuberosa oca, 100, 102

pain-killers, 143–6

paints, 126–7, 204–7
    see also dyes; inks; pigments
Palisota schweinfurthii, 151
palm, African oil *Elaeis
    guineensis,* 10, 14, 17
    food, 67, 71, 73–5
    soaps, 1–3
palm, Brazilian wax *Copernicia
    prunifera,* 24, 25, 62, 120,
    180
palm, buriti *Mauritia flexuosa,*
    34, 35, 132
palm, Chambira *Astrocaryum
    chambira,* 35
palm, coconut *Cocos nucifera,* 3,
    17, 191, 203
    coir, 3, 125, 187
    copra, 2, 3
    food, 67, 71, 73, 74, 75
    toiletries, 1, 2–3, 10, 14
palm, date, 67, 77, 87, 187
palm, raffia, 40–1
palm, sago *Metroxylon sagu,* 44,
    67
palm *Chamaerops humilis,* 120
Papaver somniferum opium
    poppy, 145–6
paper, 37, 195–202
paper mulberry *Broussonetia
    papyrifera,* 36, 37, 38, 196
papyrus *Cyperus papyrus,* 184,
    187, 195, 196
Parthenium argentatum guayule,
    173
pasta, 70
pastures, 72–3
peach nut, 8, 17
peanut/groundnut oil, 2, 17
    margarine, 71, 73
pear *Pyrus communis,* 67, 119
pecan *Carya illinoiensis,* 215
pectin, 77
pencils, 208–9
Penicillium spp., 139
Pennisetum glaucum pearl
    millet, 82
pepper *Piper nigrum,* 98
peppers *Capsicum* spp., 23, 67
pequi fruits *Caryocar brasiliense,*
    23
perfumes, 26–33, 115, 141
    toiletries, 1, 2, 7–8, 11–12, 13,
    14
periwinkle, rosy *Catharanthus
    roseus,* 159–60
persimmon *Diospyros
    virginiana,* 216, 217
petroleum nut tree *Pittosporum
    resiniferum,* 193
Phleum pratense timothy grass,
    73
Phormium tenax New Zealand
    flax, 39–40, 197
photography, 210–11
Phragmites australis, 130
Phyllostachys nigra, 219

Physostigma venenosum calabar
    bean, 150
Phytolacca bogotensis atuqsara, 6
piassava *Attalea funifera,* 128
Picea spp., *see* spruce
pigface fruits *Carpobrotus* spp.,
    97
pigments, 204–7
    see also colours, cosmetic; dyes
pigweed *Portulaca oleracea,* 97
pills, 146
pine *Pinus* spp., 108, 169, 202
    construction and joinery, 107,
    108, 112, 119, 186, 189
    dyes and paints, 58, 204
    medical usage, 138, 140, 141
    perfumes, 1, 30, 32–3
    pulp, 54, 56, 127
    southern pines, 110, 111, 200
pine, loblolly *Pinus taeda,* 200
pine, lodgepole *Pinus contorta,*
    107, 108
pine, longleaf *Pinus palustris,*
    108, 110, 200, 204
pine, Monterey *Pinus radiata,*
    198, 200
pine, parana, 113
pine, radiata, 113
pine, Scots *Pinus sylvestris,* 108,
    110, 111, 186
pine, shortleaf *Pinus echinata,*
    110, 200
pine, slash *Pinus elliotti,* 200
pine, Virginia *Pinus virginiana,*
    110
Pinus patula, 200
pineapple *Ananas comosus,* 53,
    99, 153
Pistacia lentiscus, 98
pit-pit grass *Coix lacryma-jobi,*
    36, 84
pith helmet, 63, 65
Pittosporum resiniferum
    petroleum nut tree, 193
Plantago spp. plantains, 152
Plantago afra, 152
Plantago ovata, 152
Plasmodium spp., 163
Plasmodium falciparum, 165
plywood, 111, 112, 177, 179,
    181, 182, 185
Podophyllum peltatum American
    mandrake, 160
polishes, 24, 120, 180
poplar, 199, 223
poppy, opium *Papaver
    somniferum,* 145–6
poppyseed oil, 204, 205
Porphyra umbilicalis laver bread,
    102
Portulaca oleracea pigweed, 97
potato *Solanum tuberosum,*
    44–5, 58, 67, 100–1, 202, 205
potato, sweet *Ipomoea batatas,*
    99, 192
Pourouma cecropiifolia, 151

printing, 58, 203
Prunus dulcis var. *amara* bitter almond, 17, 28
Prunus dulcis var. *dulcis* sweet almond, 17, 67
Pseudotsuga menziesii Douglas fir, 107, 109, 110, 111, 112, 186, 189
Psophocarpus tetragonolobus winged bean, 102
Pterocarpus soyauxii ngélé tree, 23
Pterogyne nitens tipa wood, 214, 215
pummelo *Citrus grandis*, 76
Pyrus communis pear, 67, 119

quebracho *Schinopsis* spp., 57, 60–1, 180
Quercus aegilops valonea oak, 61
Quercus petraea sessile/durmast oak, 118
Quercus robur pedunculate/common oak, 26, 118, 119
Quercus suber cork oak, 126, 214
Quillaja saponaria soap tree, 6
quinine, 159, 163–6
quinoa *Chenopodium quinoa*, 84

raffia, 40–1
rainforests, 56, 98–9, 112, 116, 176, 192
  clothing fibres, 34–6
  home, 131–3
  medicines, 143, 150, 151, 158–9, 161, 167
ramie *Boehmeria nivea*, 48, 49
ramin *Gonystylus bancanus*, 210, 216, 217
rape seed oil, 17, 61, 73
Raphia farinifera, 40
Raphia regalis, 40
Raphia vinifera, 40
rattans, 122–3, 213, 214
Rauvolfia serpentina snake root, 162–3
Rauvolfia vomitoria African snake root, 163
redwood, Californian *Sequoia sempervirens*, 109, 113
reeds, 130, 218–19
  totora, 182, 184
Reseda luteola weld, 58
resins, 102, 120, 131, 141, 204
  paints, 207, 217
  perfumes, 27, 29
Rhamnus purshiana buckthorn, 153
Rheum palmatum, 153
Rheum × *cultorum* rhubarb, 26
Rhodymenia palmata, 102
rhubarb *Rheum* × *cultorum*, 26
Rhus spp. sumac tree, 61, 208
Rhus verniciflua lacquer/varnish tree, 120

rice *Oryza* spp., 13, 22, 201
  food, 67, 81, 101, 133
rice, African *Oryza glaberrima*, 81
rice, Asian *Oryza sativa*, 81
rice, wild *Zizania aquatica*, 81
Ricinodendron rautanenii manketti tree, 96
Ricinus communis castor oil plant, 10
  see also castor oil
road vehicles, 169–70, 174, 176–7, 179, 180
rope, 40, 41, 49
  rigging, 186–9
Rosa canina dog rose, 31
Rosa centifolia, 31
Rosa damascena damask rose, 31
Rosa gallica, 31
Rosa odorata tea rose, 31
rosemary, 8, 21, 30
roses, 29, 30, 31, 32
  attar of, 16, 31
rosewood, *see Dalbergia* spp.
rosin, 124, 140–1, 169–70, 202, 203
rubber, 62, 140–1, 151–2, 209, 213
  tyres, 169–75
rubber tree, Ceará *Manihot glaziovii*, 173
rubber tree *Hevea brasiliensis*, 140, 169, 171, 172–3, 209
rubber tree, India *Ficus elastica*, 173
Rubia peregrina wild madder, 205
Rubia tinctorum madder, 58, 205–7
Rubus chamaemorus cloudberries, 96
Rumex arcticus sourdock, 96
rushes, 123–4
rye *Secale cereale*, 69, 87
ryegrass, Italian *Lolium multiflorum*, 72, 73
ryegrass, perennial *Lolium perenne*, 72, 73

Saccharomyces cerevisiae yeast, 69
Saccharum officinale, *see* sugar cane
safflower *Carthamus tinctorius*, 71, 73, 74, 75
saffron, 14, 16, 29
sal trees *Shorea* spp., 51, 157, 204
Salix spp. willow, 121, 181, 223
Salix alba white willow, 143, 145
  var. *caerulea* cricket bat willow, 211–12
Salix purpurea, 121
Salix triandra, 121
Salix viminalis, 121

saltwort, spiky leaved *Salsola kali*, 2
Salvadora persica toothbrush tree, 15
sandalwood *Santalum album*, 8, 22, 30, 33
sandalwood, West Indian *Amyris balsamifera*, 33
Sanguinaria canadensis bloodroot, 160
Santalum album sandalwood, 8, 22, 30, 33
sapele *Entandrophragma cylindricum*, 176, 221
Saponaria officinalis soapwort, 5
Sargassum spp., 20
satinwood *Chloroxylon swietenia*, 221
Schinopsis spp. quebracho, 180
Schinopsis balansae, 57, 60–1
Schinopsis quebracho-colorado, 60–1
Schleichera oleosa lac tree, 10
Schoenoplectus lacustris bulrush, 123
Schoenoplectus riparius totora reed, 184
Schoenoplectus tatora totora reed, 184
seaweeds, 14, 19, 127, 170, 180, 192
  cosmetics, 18, 20, 22
  food, 96, 102–3
  medical usage, 139, 146
  sizing agents, 42, 125
Secale cereale rye, 69, 87
senna *Cassia* spp., 152, 153
Sequoia spp., 112
Sequoia sempervirens Californian redwood, 109, 113
Sequoiadendron giganteum giant sequoia/Wellingtonia, 109
serendipity berry *Dioscoreophyllum cumminsii*, 80
shampoos, 7–9
shee-sham wood *Dalbergia* spp., 220
shellac, 26, 120, 208
shoes, 59–62
Shorea spp. sal trees, 51, 157, 204
silk, 50–2, 157, 181
silkworm *Bombyx mori*, 50–2
Simmondsia chinensis jojoba, 7, 8, 24
sisal, *see Agave* spp.
sizing agents, 44, 125, 202
snake root *Rauvolfia* spp., 162
soap tree *Quillaja saponaria*, 6
soaps, 1–5
  alternative, 5–6
soapwort *Saponaria officinalis*, 5
Solanum latifolium, 38
Solanum tuberosum potato, 44–5, 58, 67, 100–1, 202, 205

Sorghum bicolor sorghum, Italian whisk, 44, 77, 81, 82, 218
sorghum, sweet, 82, 192
sourdock *Rumex arcticus*, 96
soursop *Annona muricata*, 99, 102
soyabean *Glycine max*, 2, 57, 126, 150
  cosmetics, 9, 17, 18, 24
  food, 71, 73, 75, 103
Sphagnum spp., 138, 139
spices, 98
spikenard, 27, 28
Spirulina spp., 103
sponge gourds *Luffa* spp., 12–13
sport, 211–18
spruce *Picea* spp., 108, 221
  construction and joinery, 107, 108, 112, 181
  pulp, 9, 54, 127, 199
spruce, Norway *Picea abies*, 108, 110, 111
spruce, sitka *Picea sitchensis*, 107, 108, 182
spruce, western white, 108
squill *Urginea* sp., 143
starch, 44–5, 58, 202, 205
Sterculia spp., 15
Stevia rebaudiana, 80
sticking plasters, 140–1
Stipa ichu ichu, 184
Stipa tenacissima esparto grass, 62, 187, 200
straw, 62, 63, 65, 201
Streptogyna, 23
Strychnos spp., 156, 159
sugar, 77
  see also sweeteners
sugar beet *Beta vulgaris*, 10, 77, 79, 192
sugar cane *Saccharum officinale*, 10, 77–9, 99, 111, 192, 201
sugar mimic alkaloids, 167
sugar/rock maple *Acer saccharum*, 77, 78, 175–6, 221
sumac tree *Rhus* spp., 61, 208
sunflower *Helianthus annuus*, 74
sunflower seed oil, 17
  margarine, 71, 73, 75
  paints, 204, 205
sweet potato *Ipomoea batatas*, 99, 192
sweeteners, 80
Swietenia spp. mahogany, 115, 116
Swietenia macrophylla, 115
Swietenia mahagoni, 115
sycamore *Acer pseudoplatanus*, 116, 214, 217, 221, 223
Synsepalum dulcificum, 80

Tabebuia serratifolia, 160

Tachardia lacca lac insect, 26, 120, 208
talcum powder, 13
Tanacetum parthenium feverfew, 21, 147
tannins, 26, 59–61, 179–80
Taraxacum bicorne dandelion, 173
Taraxacum officinale dandelion, 58, 87
taro *Colocasia esculenta*, 37, 38, 99
  var. *antiquorum*, 44
Taxus baccata yew, 119
tea, 89–95
tea, Assam *Camellia sinensis* var. *assamica*, 92
tea, China *Camellia sinensis* var. *sinensis*, 92
tea, labrador *Ledum palustre*, 96
teak *Tectona grandis*, 111, 113, 114, 116, 176, 185, 186
t'ef *Eragrostis tef*, 84
teosinte *Euchlaena mexicana*, 83
Terminalia chebula myrabolans tree, 61
Terminalia ferdinandiana billygoat plum, 97
Thaumatococcus danielli, 80
Thea assamica, 92
Theobroma cacao cocoa, 18, 98
thickeners,
  cosmetics and toiletries, 8–9,

14, 18, 22
  dyes and inks, 45, 58
thrombosis, 153
Thuja plicata Western red cedar, 112, 113
thyme, 29, 143
Tilia americana basswood, 221
Tilia europaea lime, 119
timber, 105–19, 184–6
  cladding, 112–13
  terminology, 110–12
tipa wood *Pterogyne nitens*, 214, 215
tobacco, 160
toothbrush tree *Salvadora persica*, 15
toothpaste, 13–15
  see also chewing sticks
toquilla *Carludovica palmata*, 65
trains, 174–6
Trema micrantha, 198
Trifolium spp. clovers, 73
Triplochiton scleroxylon obeche, 131, 185, 214
Tripsacum spp., 83
Triticum aestivum bread wheat, 70
Triticum durum durum wheat, 70
Triticum turgidum English wheat, 70
Tropaeolum tuberosum aiñu, 102
Tsuga spp. hemlock trees, 61

Tsuga heterophylla western hemlock, 9, 56, 109, 110, 112
tulip tree *Dalbergia cearensis*, 181
tung oils *Aleurites* spp., 126, 203
turmeric *Curcuma longa*, 38
turpentine, 1, 32, 33, 127, 141, 204–5
Turraeanthus africanus asama bark, 151
tyres, rubber, 169–75

Ullucus tuberosus ulluco, 102
Urginea sp. squill, 143
Urtica dioica nettle, 8, 21, 48, 58

Vaccinium myrtillus bilberry, 96
valerian, Himalayan *Nardostachys jatamansi*, 27
vanilla plant *Vanilla planifolia*, 14
vitamins, 69, 76, 77, 96, 97, 101–2

wallpaper, 127–8
walnut, 116, 176, 178–9, 214
  oil, 204
walnut, black *Juglans nigra*, 26, 178–9
walnut, English *Juglans regia*, 178, 179
wattle, *see Acacia* spp.

wax, candellila, 24, 62, 208
wax, carnauba, 24, 25, 62, 120, 180, 208
waxes, 24, 62, 120, 180, 208
weld *Reseda luteola*, 58
Wellingtonia *Sequoiadendron giganteum*, 109
wheat *Triticum* spp., 70
  food, 67, 70, 81, 101
  starch, 13, 44, 58
  straw, 63, 65
wickerwork, 121, 181
willow, *see Salix* spp.
woad *Isatis tinctoria*, 58
wood pulp, 15, 68
  paper making, 127, 195, 198–200
  synthetic fibres, 54, 55, 56
  thickeners, 8–9, 13
wych hazel *Hamamelis virginiana*, 140

yams *Dioscorea* spp., 99, 149–50
yeast *Saccharomyces cerevisiae*, 69
yew *Taxus baccata*, 119
ylang-ylang *Artabotrys odoratissimus/Cananga odorata*, 30, 31

Zea diploperennis, 83
Zea mays, *see* maize
Zizania aquatica wild rice, 81

Author's acknowledgements

The following companies and organizations were consulted whilst researching this book

Ace Kensington
A.E. Olney & Co. Ltd.
African Territories Wattle
    Industry Fund
Agfa Gevaert Ltd.
All-Sport International
Alvis Ltd.
Amateur Photographer
Aqualon UK Ltd.
Association of Jute Spinners &
    Manufacturers
Barbara Attenborough
    Associates
Bayley's Caravans
B.D.H. Chemicals
Belmont Weaving
Bill of Entry (Statistical Office)
The Body Shop International
    PLC
Bodytreats Ltd.
J.W. Bollum & Co. Ltd.
Boot & Shoe Manufacturer's
    Association
Boots PLC
Boulton & Paul (Joinery) Ltd.
Bowater Paper Company
Bridon Fibres Ltd.
Brightstern Ltd.
Britcair
British Aerospace
British Airways
British Bakeries Ltd.
British Brush Manufacturers'
    Association
British Carpet Manufacturers'
    Association
British Crown Green Bowling
    Association
British Footwear
    Manufacturers' Federation
British Institute of Professional
    Photography
British Leather Confederation
British Leather Co. Ltd.
British Leather Research
    Institute
British Rail
British Rope
British Rubber Manufacturers'
    Association Ltd.
Buckingham Palace Press Office
Bush Boake Allen Ltd.
California Cedar Products
    Company
Castrol U.K. Ltd.
Central Chemicals Co. Ltd.
Centre for Rheumatic Disease,
    Glasgow Royal Infirmary

Chamber, Cox & Co.
Charcoal Cloth Ltd.
Chelsea Physic Garden
Chesham Chemicals Ltd.
Chipboard Promotion
    Association
Ciba Geigy
Clipston & Whitwell
The Cordage Manufacturers'
    Institute
Council of Forest Industries of
    British Columbia
Country Chairmen
Courtaulds Chemicals
Courtaulds PLC
Courtaulds Pulp Group
    Services
C.P.C. (UK) Ltd.
C.S.E. Aviation
Cumberland Pencil Co. Ltd.
Cunard
Doughty Group
Duckers Furniture Ltd.
Duncan, Flockhart & Co. Ltd.
Dunlop Aviation Tyres Ltd.
Dunlop Slazenger International
    Ltd.
E.C. De Witt & Co. Ltd.
Elastoplast
Elida Gibbs Ltd.
Embassy of the German
    Democratic Republic
Embassy of the Philippines
English Bowling Association
Ethicon Ltd.
Federation of Aromatherapists
Federation of Oils, Seeds and
    Fats Associations Ltd.
Ferguson & Menzies Ltd.
Fidor (Fibre Building Board
    Association)
Firmenich
The Flour Advisory Bureau
Flour Milling & Baking
    Research Association
Food & Drink Federation
Footwear Technology Centre
Friends of the Earth
G & G Food Supplies
Gladsby & Son
Goodyear Tyre & Rubber Co.
    GB Ltd.
Gover, Horowitz & Blunt Ltd.
Greenwich Maritime Museum
Gunn & Moore Ltd.
Haarman & Reimar
Habitat Designs Ltd.
Hallite Seals International

Hancock & Roberts Ltd.
Hardwood Dimensions Ltd.
Hattersley's Lacrosse Works
Heliks International
Henkel Chemicals Ltd.
Hercules Ltd.
Hollins Paper Mill
Horniman Museum & Library
ICI (Paints Division)
Interbed
International Coffee
    Organization
International Linen
International Paper Company
    Ltd.
Jacobs, Young & Westbury
Japanese Embassy
Jewson Ltd.
John & Arthur Beare
Johnson & Johnson Ltd.
Johnson Wax
Kodak Ltd.
Lawrie Bagpipes Ltd.
Leeds University Department
    of Textile Industries
Lever Brothers Ltd.
Bill Lewington's Musical
    Instruments
Liverpool Museum
London Rubber Company
    Products Ltd.
London Underground Ltd.
The Malaysian Rubber
    Producers' Research
    Association
Mary Rose Trust
Master Rope Makers Ltd.
Merseyside Maritime Museum
Metro Cammell
Ministry of Agriculture, Food
    & Fisheries
Morgan Motor Co. Ltd.
Morris Minor Owners' Club
Morris Traveller Centre
Museum of Mankind
NASA HQ
National Bakery School
National Footwear Museum
Neils Yard Apothecary
New Zealand High
    Commission
Overseal Ltd.
The Paint Research
    Association
P & O
Peaudouce (UK) Ltd.
Pelikan
Pharmaceutical Society of

Great Britain
PIRA (Paper Industries
    Research Association)
Pittard Garner
Pothe Hille
Price Pierce
The Proprietary Association of
    Great Britain
Reckitt & Coleman
Reed Paper & Board (UK) Ltd.
Rexel
Rio Trading Company (Health)
    Ltd.
Rockwell International
Rolls-Royce Motor Cars Ltd.
Romany Herb Products Ltd.
Royal Aeronautical Society
R.S. Lawrence
Saab GB Ltd.
Arthur Sanderson
Scott Lithcow
The Seed Crushers' & Oil
    Producers' Association
Shell Lubricants UK
Shirley Institute
Shoe & Allied Trade Research
    Association
Silk Association of Great
    Britain
Smith & Nephew Medical Ltd.
Soap, Perfumery & Cosmetics
Speciality Pulp Services Ltd.
Spicers Board Mill
S.P. Tyres UK Ltd.
Staedtler UK Ltd.
Survival International
Susanna Wood
T & A Wardle
Tanning Extract Producers'
    Federation
The Tea Council
Trada (Timber Research &
    Development Association)
Treforest
Tunnel Avebe Starches Ltd.
The United Kingdom & Ireland
    Particleboard Association
The Weald & Downland
    Museum
Wellcome Institute
W.H. Bence Coachworld
W.H. Smith & Son
Wiggins Teape
Wimpey Homes Holdings Ltd.
Windsor & Newton
Worldwide Butterflies &
    Lullingstone Silk Farm
Yamaha Kemball Ltd.

I gratefully acknowledge the help of the following individuals

Phil Agland	Janette Eardley	Dr Mike Hitchcock	Mr & Mrs Alistair McLeod
Bryan & Cherry Alexander	Dr John Edmondson	Tim Holden	Mathews
Vernon S. Bach	John Fallon	Roger & Peter Hunt	David Mitchell
Dr Les Blackwell	Dr Linda Fellows	Jay Kettle Williams	Keith Nicklin
Gillian & Haydon Bradshaw	Jim Forrester	Mr William King	John Stickland
Mr Peter Bridle	Conrad Gorinsky	Dr Neil Kitchen	Henry Stobbart
Rosemary Davies	Sue Grierson	Andrew Lee	Mr Michael Taylor
David Dowse	Dr Michael Harrison		

Special thanks to Dr Brinsley Burbidge (Head of the Information & Exhibitions Division at the Royal Botanic Gardens, Kew) at whose invitation this book was written, all the other members of the Kew staff who not only helped me but made me very welcome whilst I researched *Plants for people*, and also to the editorial team of the Publications Department of The Natural History Museum.

My warmest thanks go to my family and to Edward Parker, whose constant encouragement and generous support helped me, beyond measure, to research and write this book.

Picture credits

All botanical drawings are from the Botany Library of The Natural History Museum, London

STARTING THE DAY

Adrian Essential Oils p.11 bottom, p.32 bottom
Ardea, London p.7 bottom (François Gohier)
Arrowmed Limited p.15
H. Bronnley & Co. p.4 right
Compix/Commonwealth Institute p.20 bottom, p.22 centre, p.25 bottom
Cosmetics to go p.32 left
Country Living Facing p.1 (Tim Beddow) p.28 right (Susan Witney)
Crabtree and Evelyn p.4 top left, p.7 top, p.9, p.13
Eric Dent p.26 top
DJA River Films p.23 bottom (Lisa Silcock)
The Face Station p.21 (John Downs, The Natural History Museum)
David George p.19 centre
Elida Gibbs p.4 bottom left
Susan Griggs Agency p.22 right (Alain Evrard)
Hennara Products p.25 top
Hutchison Library p.3 top (Patricio Goycolea) p.16 (Dr Nigel Smith) p.22 left (S. Errington) p.23 centre (John Wright)
Dr Edmund Launert p.26 bottom, p.28 left
Norfolk Lavender Farm p.11 top right
Edward Parker p.3 bottom left & right, p.6, p.29
Jim Price p.19 bottom
Scarborough & Company p.1
Survival International p.23 top (Victor Engelbert)

KEEPING US COVERED

British Museum (Natural History) p.39
Compix/Commonwealth Institute p.36, p.42 left (Reg Varney) p.52 left (M. Proctor)
Dollond & Aitchison p.57
Girl About Town/Red or Dead p.61
Hutchison Library p.35 (Michael Macintyre) p.37 (Patricio Goycolea)
Irish Linen Company p.47
Japan National Tourist Organisation p.52 right
Luton Museums and Galleries p.64
Monsoon p.41
Mothercare p.48
Director, National Army Museum p.63 bottom
Edward Parker p.40, p.42 right, p.45, p.53 top
Peaudouce p.56
Royal Botanic Gardens, Kew p.38, p.49, p.53 bottom
Glyn Satterley p.59
Jane Smith Straw Hats p.63 top
Gillian Spires p.58
Tun Abdul Razak Laboratory p.62
Warehouse p.55 bottom
Wiggins Teape p.55 top (Colin Molyneux)
Worldwide Butterflies Ltd. p.51

FROM FIRST FOODS TO FAST FOODS

Ardea, London p.99 (Don Hadden) pp.102, 103 (John Mason)
Bryan and Cherry Alexander Photography p.96
Heather Angel p.97
British Sugar Bureau p.79 right
Compix/Commonwealth Institute p.70 top, p.78, p.79 left, p.81
Crabtree & Evelyn p.94
Dorling Kindersley p.66 (Charlie Stebbings) p.67 (Martin Brigdale) p.69 (Clive Streeter) p.70 bottom (Ian O'Leary) p.98 (Philip Dowell)
International Coffee Organisation p.84, p.85, p.88
Japan National Tourist Organisation p.90
Edward Parker p.71, p.72, p.75, p.83, p.100
J. Sainsbury PLC inset p.66
Tate & Lyle p.80
The Tea Council p.89, p.91, p.93, p.95

PLANTS THAT PROTECT US

Heather Angel p.108 top, p.109 centre & bottom, p.122 left
Ardea, London p.108 bottom (François Gohier) p.126 (Jack A. Bailey)
Mark Boulton/ICCE p.116
Trustees of the British Museum © 1988 p.133 bottom
Coloroll Home Furnishings Ltd. p.113
Compix/Commonwealth Institute p.110, p.111, p.125 right
Eric Dent p.106 right
DJA River Films p.131 top (Lisa Silcock)
Fencepost Prints, Brisbane p.112 right (Trish Ferrier)
M J Givans p.115 bottom, p.129
Goethe Institute p.106 left
Gostin of Liverpool Ltd. p.115 top
Grange Furniture UK p.122 right
Guildway Timber Ltd. p.107 bottom
Habitat Designs Ltd. p.125 left, p.127 left
Hutchison Library p.107 top (Leslie McIntyre) p.117 bottom (Philip Wolmuth) p.130 (V. Southwell) p.133 top right (Brian Moser)
ICI Paints p.127 right
Rob Judges p.123
Marquess of Bath p.114 top
The National Maritime Museum p.114 bottom, p.118 bottom
Edward Parker p.104, p.108 centre, p.117 top, p.118 top, p.131 bottom
Royal Botanic Gardens, Kew p.109 top
M L Smith p.112 left
South American Pictures p.133 top left (Bill Leimbach)
Gillian Spires p.121
Woodstock Furniture p.119

PLANTS THAT CURE US

Heather Angel p. 148, p.150, p.161
Ardea, London p.147 top, p.153 (J L Mason) p.162 (Kenneth W Fink)
DJA River Films p.147 bottom right, p.151
Alan Eddy p.138 bottom
Gerard House Ltd. p.144
M J Givans p.140
Hutchison Library p.155 (Brian Moser) p.156 (John Wright)
Oxford Scientific Films p.145 (Scott Carnazine)
Edward Parker p.134, p.137, p.138 top, p.143, p.147 bottom left, p.149, p.157, p.158, p.163, p.166, p.167
A V Pound & Co Ltd. p.141
Steriseal p.139

PLANTS THAT TRANSPORT US

Heather Angel, p.179 bottom
Ardea, London p.186 right, p.189 (Nick Gordon)
Bridon PLC p.187
British Airways PLC p.170 left
British Museum (Natural History) p.177 left (Colin Keates)
The Caravan Club p.177 right
Castrol UK Ltd. p.192
Compix/Commonwealth Institute p.183 top & bottom
Goodyear GB Ltd. p.172
Japan National Tourist Organisation p.174
London Transport Museum p.175
Edward Parker p.168, p.170 centre, p.182, p.183 centre, p.185 left, p.186 left, p.190
Rolls-Royce Ltd. p.179 top
Science Museum, London p.180, p.181 bottom
The Scottish Fisheries Museum Trust Ltd. p.185 right
Taylor Photographic Library p.181 top
Tun Abdul Razak Laboratories p.170 right, p.171, p.173

PLANTS THAT ENTERTAIN US

Action-Plus Photographic p.125 top left, top right (Lee Wadle) p.218 (Mike Hewitt)
Heather Angel p.215 bottom
Ardea, London p.196 (Peter Steyn) p.200 (A P Paterson) p.205 top (John Mason) p.210 top
The Belmont Press p.202 left
Billiards and Snooker Archive Collection p.216 centre
Bowls International p.217 (Duncan Cubitt)
British Museum (Natural History) p.204 (Colin Keates)
Paul Bryant p.222
California Cedar Products Co. p.208 bottom
Daler Rowney Ltd. p.205 bottom, p.208 top
Miguel Fairbanks p.209 left
Guardian Newspapers p.202 right
Gunn & Moore p.211 & 213
Hutchison Library p.195 (Bernard Regent) p.206 (Jeremy Horner) p.215 centre (Alan Hutchison)
Japan National Tourist Organisation p.219 right
Kodak Ltd. p.210 bottom
Mexicolore p.198 (Sean Sprague)
Edward Parker p.219 left
Peradon & Fletcher p.216 bottom
Robert Horne p.194
Royal Philharmonic Orchestra p.221
St. Bride's Printing Library p.201
William Sinclair & Son p.220
Gillian & David Spires p.197 top series of pictures
Tun Abdul Razak Laboratories p.209 right
UK Paper p.197 bottom right
Wiggins Teape p.199
Yamaha-Kemball Music p.223